Platelets

Edited by Steve W. Kerrigan

Published in London, United Kingdom

IntechOpen

Supporting open minds since 2005

Platelets
http://dx.doi.org/10.5772/intechopen.77663
Edited by Steve W. Kerrigan

Contributors
Andaleb Kholmukhamedov, Lucy Kornblith, Zachary Mattthay, Alper Saglanmak, Caglar Cinar, Muhammet Mesut Nezir Engin, Raul Teruel-Montoya, Ernesto Cuenca-Zamora, Constantino Martínez, Sonia Aguila, Bela Balint, Mirjana Pavlovic, Milena Todorovic, Mrinal Kanti Poddar, Soumyabrata Banerjee, Bernhard Resch, Alper Gultekin

First published in London, United Kingdom, 2020 by IntechOpen
IntechOpen is the global imprint of INTECHOPEN LIMITED, registered in England and Wales, registration number: 11086078, 5 Princes Gate Court, London, SW7 2QJ, United Kingdom
Printed in Croatia

British Library Cataloguing-in-Publication Data
A catalogue record for this book is available from the British Library

Additional hard and PDF copies can be obtained from orders@intechopen.com

Platelets
Edited by Steve W. Kerrigan
p. cm.
Print ISBN 978-1-83881-114-3
Online ISBN 978-1-83881-115-0
eBook (PDF) ISBN 978-1-83881-116-7

For EU product safety concerns:
IN TECH d.o.o., Prolaz Marije Krucifikse Kozulić 3, 51000 Rijeka, Croatia, info@intechopen.com or visit our website at intechopen.com.

Meet the editor

Professor Steven W. Kerrigan is deputy head of the School of Pharmacy (Research), head of the Cardiovascular Infection Research Group at the Royal College of Surgeons in Ireland (RCSI), University of Medicine and Health Sciences, and inventor of the sepsis treatment drug InnovoSep. Professor Kerrigan is a graduate of King's College London, England (Pharmacology), University of Strathclyde, Scotland (Immunopharmacology), and RCSI (Infection and Immunity). His research focuses on understanding the platelet and endothelial response to infection during sepsis. Through research, Professor Kerrigan identified a promising drug target that prevents a wide number of microorganisms (bacteria, fungus, and virus) from causing a dysregulated response in the systemic circulation during sepsis, specifically preventing unwanted platelet and endothelial cell activation. Professor Kerrigan has published extensively in leading high-impact journals in the areas of platelets, endothelial cells, and bloodstream infections, and has attracted more than €6.5 million in grant funding and filed three patent/disclosures. Professor Kerrigan is currently co-chair of the ISTH Scientific Standardization Committee Biorheology (platelets) and member of the European Sepsis Alliance research committee.

Contents

Preface

Mammalian platelets are small (2–4 um), discoid, short-lived fragments derived from megakaryocyte precursors. They play a crucial role not only in the formation of a normal hemostatic plug but also in the formation of a pathologic thrombus, particularly within arteries subjected to high shear stress. Platelets circulate in a resting state through blood vessels lined by an endothelial layer. Following an insult to the endothelial layer platelets undergo a series of changes that ultimately lead to them changing shape, secreting their granule contents, and aggregating to prevent blood loss. Thus, platelets rely on specific adhesive interactions with extracellular matrices (e.g., collagen, von Willebrand factor, fibronectin). In order to achieve firm adhesion, platelets possess many receptors specific for extracellular matrix proteins. As our basic knowledge in platelet biology has improved through the years, it is has become evident that the role of platelets extends beyond that of thrombosis and hemostasis. Current state of the art suggests that platelets play a key role in a much wider repertoire of physiological processes such as inflammation, innate immunity, cancer, infection, neurobiology, and tissue repair/regeneration.

The individual chapters in this book identify one particular aspect of platelet function, dysfunction, or application. The book is not intended to be read from cover to cover as the reader seeks authoritative information on any one given aspect of platelets. Therefore, for ease of reading, the book is divided into three main sections: "Platelet Function" and "Platelet Dysfunction" and "Platelet Application." The authors of each of the focused chapters are leaders in their respective fields. As significant advances continue to develop our thinking of the functional role of platelets in health and disease, this book elevates awareness and enthusiasm in further investigating platelet functions.

Steven W. Kerrigan
Professor,
Head of the Cardiovascular Infection Research Group,
RCSI University of Medicine and Health Sciences,
Dublin, Ireland

Section 1

Platelet Function

Chapter 1

Platelet Imaging

Zachary A. Matthay and Lucy Zumwinkle Kornblith

Abstract

The knowledge gained through imaging platelets has formed the backbone of our understanding of their biology in health and disease. Early investigators relied on conventional light microscopy with limited resolution and were primarily able to identify the presence and basic morphology of platelets. The advent of high resolution technologies, in particular, electron microscopy, accelerated our understanding of the dynamics of platelet ultrastructure dramatically. Further refinements and improvements in our ability to localize and reliably identify platelet structures have included the use of immune-labeling techniques, correlative-fluorescence light and electron microscopy, and super-resolution microscopies. More recently, the expanded development and application of intravital microscopy in animal models has enhanced our knowledge of platelet functions and thrombus formation *in vivo*, as these experimental systems most closely replicate native biological environments. Emerging improvements in our ability to characterize platelets at the ultrastructural and organelle levels include the use of platelet cryogenic electron tomography with quantitative, unbiased imaging analysis, and the ability to genetically label platelet features with electron dense markers for analysis by electron microscopy.

Keywords: platelets, platelet disorders, electron microscopy, confocal microscopy, fluorescence microscopy, intravital microscopy

1. Introduction

Platelets are small, anucleate cells with central roles in hemostasis, endothelial regulation, inflammation, and immune responses [1–5]. Critical to these biologic processes, platelets undergo several dynamic morphologic changes in overall shape and size. Furthermore, at the organelle level they alter and differentially release their cellular contents in signal dependent manners [6–10]. Detailed knowledge of these changes is essential for understanding platelet biology, including in the contexts of hemostasis and thrombosis, cellular and pathogen interactions, and intracellular signaling mechanisms [9, 11–14]. Within this chapter, we will review the methods and applications of multiple modalities of platelet imaging including light transmission, super-resolution, electron, and intravital microscopy. Advances in our ability to image platelets have enabled a more complete understanding of platelet biology in health and disease, and these techniques remain powerful clinical and research tools [15–19]. We will also detail the challenges of these imaging methods specific to platelets, and future barriers that should to be addressed.

2. Light and fluorescence microscopy

2.1 Overview

Owing to their small size, platelets were not discovered until long after red blood cells and leukocytes had been visualized with traditional microscopy. With the development of compound and achromatic microscopes, they were finally observed by several scientists in the nineteenth century, including the Italian scientist Giulio Bizzozero, who began to identify their roles in hemostasis and thrombosis [14, 20]. In their resting state, platelets are 2–3 μm in diameter and discoid in shape, appearing as small blue circular structures on traditional light microscopy with Giemsa staining (**Figure 1**). Despite the resolution limitations of light microscopy, many important advances have been made in our understanding of platelet structure and function with this technique. In fact, light microscopy remains an important diagnostic and research tool today [21, 22].

2.2 Diagnosis of platelet disorders with light and fluorescence microscopy

Many heritable platelet disorders can be diagnosed using light microscopy of blood smears when used in conjunction with immunofluorescent staining. Typically, light microscopy is initially used to identify abnormalities in platelet shape and size, narrowing the differential diagnosis, which is then followed by specific staining or flow cytometry to make a final diagnosis (**Figure 2**) [21]. This may be diagnostic in approximately 25–35% of cases of suspected inherited thrombocytopenia, and can obviate the need for further expensive genetic testing or complex imaging modalities such as electron microscopy [21, 22].

2.3 Platelet spreading assays

Differential interference contrast microscopy is a specialized form of light microscopy used to characterize platelet spreading and adhesion to proteins on transparent surfaces [24]. For example, the role of actin proteins in the platelet

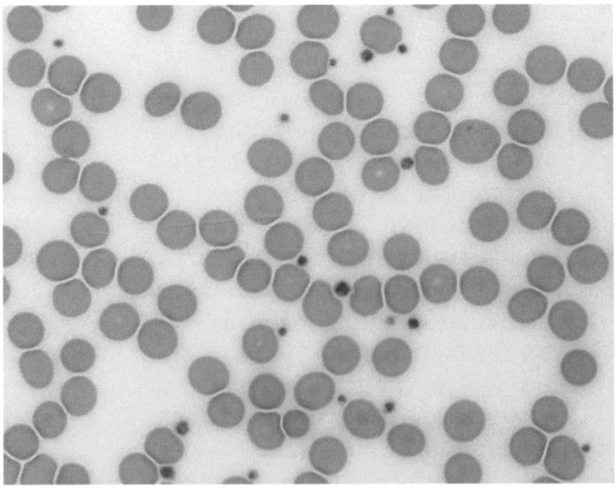

Figure 1.
Standard light microscopy of platelets. Image from a light microscope (500×) from a Giemsa-stained peripheral blood smear showing platelets (blue dots) surrounded by red blood cells (pink circular structures). Image and caption by Dr. Graham Beards, reproduced under Creative Commons GNU Free Documentation 1.0 Generic license (https://creativecommons.org/licenses/by-sa/1.0/deed.en).

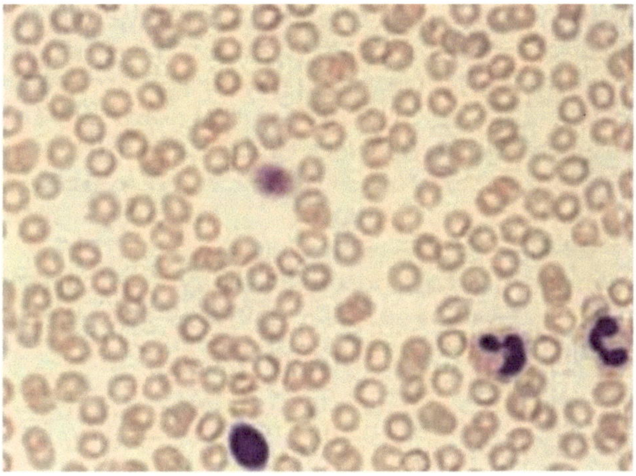

Figure 2.
Light microscopy of peripheral blood smear. Performed with Giemsa staining, showing a giant platelet characteristic of Bernard-Soulier syndrome. Diagnosis can then be confirmed by either immune-fluorescent microscopy or flow cytometry demonstrating defects in glycoprotein Ib. Image reproduced with permission from Sandrock et al. [23].

responses to collagen or fibrinogen can be studied (**Figure 3**) [25]. This form of microscopy enhances contrast in unstained platelet samples thereby allowing adequate visualization of shape change [24]. Platelet spreading experiments have led to several important advances in platelet biology such as the identification of sex differences in platelet adhesion [26], discovery of novel platelet structures such as actin nodules [27], and delineation of platelet-extracellular matrix interaction [25, 28].

2.4 Imaging platelets under flow

In order to study platelet behavior in conditions that more closely replicate human vascular biology, techniques that incorporate the imaging of platelets under flow conditions have been developed [8, 29–31]. These include *in vitro* methods such as "flow chambers" and *in vivo* intravital microscopy (detailed in Section 5). Flow chambers include a wide range of home-made and commercially available

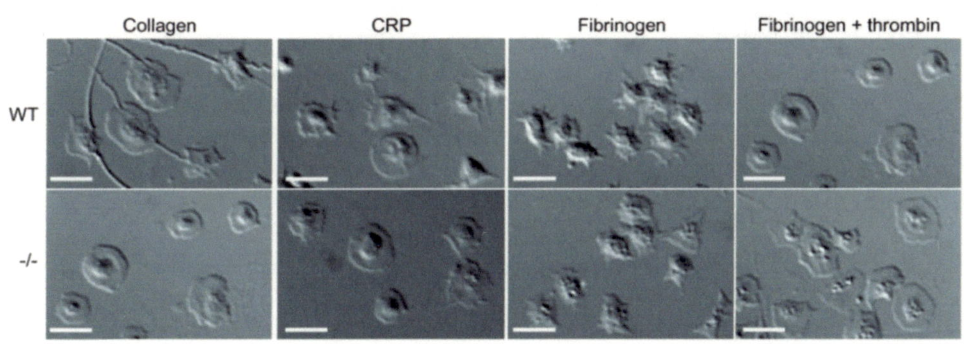

Figure 3.
Analysis of platelet spreading and actin organization. Washed platelets (2 × 10^7 platelets/ml) from wild type and HS1 (actin binding protein) −/− mice were added to cover slips coated with collagen (100 µg/ml), collagen related peptide (10 µg/ml) or fibrinogen (100 µg/ml) ± thrombin (1 U/ml) and allowed to settle for 45 or 90 min at 37°C. Spread platelets were fixed in formalin and imaged using differential interference contrast microscopy. Representative images of platelets at 45 min are shown. Image and caption reproduced with permission from Thomas et al. [25] under Creative Commons Attribution License (http://creativecommons.org/licenses/by/2.0).

devices such as parallel-plate flow chambers or biochips with micro channels through which fluid is passed at arterial or venous shear stresses [32, 33]. Confocal and fluorescence microscopy is then used to visualize platelet adhesion, aggregation and thrombus formation [30, 33]. Investigators can create a range of experimental conditions to study platelet behavior which has led to many advances in our understanding of platelet biology, thrombus formation, and responses to antiplatelet therapies [29, 32]. Most recently, for example, Tunstromer et al. used platelet staining and time-lapse microscopy to track the movements of individual platelets during thrombus formation using a flow chamber [29]. This model was applied to quantitatively study the effects of platelet inhibitors on platelet contraction, revealing differential impacts on platelet subpopulations [29].

3. Super-resolution microscopy

3.1 Overview

Standard light microscopy offers a maximum resolution of about 200–300 nm due to the limits on resolution to about half of the wavelength of visible light [34]. This allows platelets to be identified, but it is difficult to distinguish and examine intracellular structures. However, recent advances in fluorescence light microscopy with the development of super-resolution microscopy techniques have improved upon conventional light microscopy resolution by 2–10 fold, to the 10–200 nm range, and the pioneers of these advances were awarded Nobel prize in chemistry in 2014 [34, 35]. These novel techniques all utilize fluorescent labeling but broadly apply three different imaging techniques (wide-field microscopy, total internal reflection fluorescence, or confocal microscopy) to overcome the resolution limit of light microscopy, with resolutions at the 10–200 nm range [36, 37]. Further, they can be coupled to automated image analysis to systematically evaluate platelet granules and reveal three dimensional structural details [34, 37].

3.2 Localization microscopy or pointillist imaging by single-molecule localization (SMLM)

Localization microscopy or pointillist imaging by single-molecule localization (SMLM) entails using many cycles of detecting and localizing single fluorescent labels within the cell achieving up to 10 nm resolution. By repeating the fluorescent imaging cycles thousands of times in conjunction with an on/off mechanism for fluorescence, the positions of molecules are precisely mapped out [37]. However, this process is quite slow given the need for several cycles of images to be taken. Sub-variants include photoactivation localization microscopy (PALM) and stochastic optical reconstruction microscopy (STORM) [34].

3.3 Target based inhibition of fluorescence emission by stimulated emission

Target based inhibition of fluorescence emission by stimulated emission uses an excitation and a depletion laser beam to manipulate fluorescent signals to further improve resolution, relying on diffraction properties of light from the microscope's focal point [37, 38]. Some examples of this include stimulated emission depletion (STED), reversible saturable optical fluorescence transitions (RESOLFT), and conical diffraction [34]. This method requires a narrow field and imaging an entire sample is a lengthy process, though the resolution is excellent at up to 20 nm [34].

3.4 Structured illumination microscopy

Structured illumination microscopy only improves resolution by approximately 2 fold (up to 120 nm) compared to light microscopy, but this is a more efficient, faster method [35]. It employs mathematical processing of a series of illumination and laser interference patterns to improve resolution [37]. As a proof of concept, Westmoreland and colleagues used structured illumination microscopy to discriminate between three patients with Hermansky-Pudlak Syndrome (a platelet storage defect whereby dense granules are absent) and healthy controls [35]. This work highlights some of the inherent limitations of super-resolution microscopy. Because it relies upon fluorescent labeling of structures of interest, there are risks of mislabeling and therefore misidentification of structures. For example, in their paper, the authors noted that in order to stain dense granules fluorescently, a CD63 marker was used, however, off target effects such as identification of lysosomes were seen as these are also known to be CD63 positive. The description of platelet actin nodules required for platelet-platelet interactions and their potential relevance to Wiskot-Aldrich syndrome is another advance that was made using structured illumination microscopy [39].

In summary, super-resolution microscopy is a recent and notable advance over traditional light microscopy techniques in the identification and visualization of platelet ultrastructural features. Despite some limitations, it does offer the advantage, particularly for structured illumination microscopy, of being relatively less time consuming and intensive than other advanced imaging techniques (such as electron microscopy), and as such may ultimately become more widely available.

4. Electron microscopy of platelets: transmission, scanning, and cryogenic

4.1 Overview of electron microscopy

Electron microscopy (EM) utilizes a beam of accelerated electrons as the source of illumination, and because the wavelength of an electron is up to 100,000 times shorter than that of visible light, this provides resolution at the nanometer level with excellent visualization of platelet ultrastructural details (see **Figure 4**). EM was first developed in the 1930s, and much of our detailed knowledge of platelets has been gleaned through application of EM, including through transmission, scanning, and cryogenic EM [40, 41]. Substantial early contributions were made by James White who analyzed platelet ultrastructure in health and disease states using EM [13, 14, 16, 42–48], and by Marcel Bessis, who studied both megakaryocyte and platelet ultrastructure in addition to other blood cell morphologies [49, 50]. In this section, we will explore the applications of EM in clinical and research settings and discuss their respective advantages and drawbacks.

4.2 Scanning and transmission electron microscopy

Scanning (SEM) and transmission electron microscopy (TEM) can magnify up to a resolution of 0.2 nm [15]. In SEM, detailed topographical images are obtained of the platelet surface. However, to examine the contents of platelets, TEM is used which sends a beam of electrons through ultrathin sections, providing excellent internal details. EM relies on differences in electron density to differentiate individual structures; however, some structures may have similar electron densities. For example, the identification and study of platelet lysosomes with EM requires

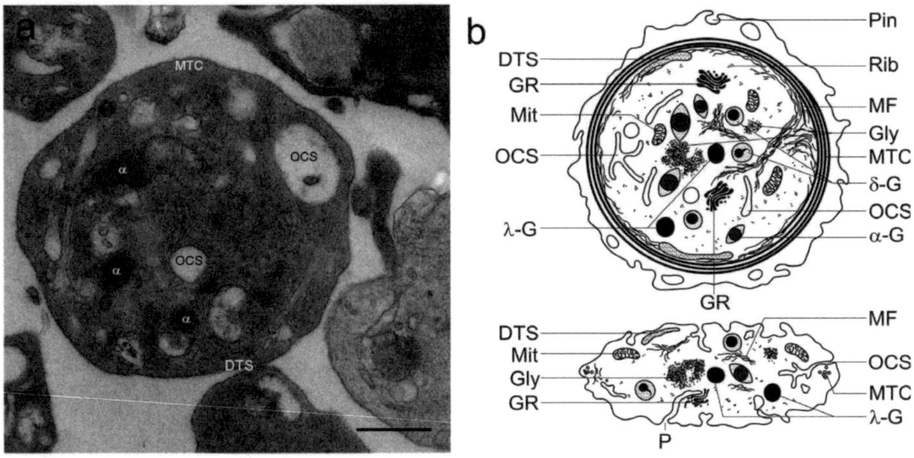

Figure 4.
Resting platelet shape and structural components. (a) Transmission electron microscopy in the equatorial plane of a resting, discoid platelet, with showing the peripheral MTC in the equatorial plane. (b) A scheme of a platelet in the equatorial plane (upper image) and in cross section (image at the bottom). Abbreviations: *MTC, microtubular coil DTS dense tubular system; Gly, glycogen; α, α-granules; δ, δ-granules or dense bodies; λ, λ-granules or lysosomes; GR, Golgi remnants; MF, microfilaments; Mit, mitochondria; OCS, open canalicular system; P, pores of the OCS; Rib, ribosomes. Figure and caption reproduced from Neumüller et al. [51], under open access Creative Commons 3.0 Licensing (https://creativecommons.org/licenses/by-sa/3.0/deed.en).*

staining with acid phosphatase or by using arylsulfatase reactive granules to differentiate them from alpha granules [52]. EM has also been used to demonstrate how platelets may release organelles encapsulated in vesicles, such as mitochondria, to drive pathologic inflammation [53, 54].

Despite the significant advances SEM and TEM contributed to the understanding of platelet structure and biology, they have important limitations. Both SEM and TEM require extensive processing of the samples. This includes fixation, dehydration, hardening or embedding, as well as sectioning. Given that platelets are prone to activation, degranulation, and microparticle release with even modest environmental stimuli, the concern is that these fixation processes risk significantly altering and activating platelets. The most common fixative used for SEM and TEM of platelets is glutaraldehyde, which preserves morphology best of the agents used. Despite these limitations, SEM and TEM have significantly contributed to our understanding of the intracellular organization of platelet contents with nanometer resolutions. Critically, SEM and TEM can be combined with immune-gold labeling to further localize structures [7]. For example, this labeling approach has been used to delineate the intracellular distribution of vWF [10], and to demonstrate that thrombin but not adenosine diphosphate treated platelets have upregulated GLUT-3 receptors [55].

4.3 Cryogenic-electron microscopy

Cryogenic electron microscopy (Cryo-EM), developed in the 1980s, is a technique in which EM is applied to samples which are nearly instantaneously cooled to cryogenic temperatures in aqueous solution [56]. This process (typically done by plunging into liquid ethane) occurs so rapidly that water molecules do not rearrange into crystalline form (also known as amorphous or vitreous ice). This preserves cellular structures such as lipid membranes in their near native states [41]. This is a particularly important advantage over TEM and SEM which require extensive fixation as described above because platelets are known to activate with minimal stimuli. Furthermore, samples can then be stored for long periods of time in liquid nitrogen. Once cryogenically preserved, standard TEM techniques are applied to the samples. Additionally, multiple "tilt-series" or tomograms are obtained, in which images are captured at multiple angles, which can then be built into 3D reconstructions (also called cryogenic electron tomography or cryo-ET), which can detail internal structures such as the open canalicular system, organelles, and granules [57, 58] (**Figure** 5).

4.4 Diagnostic and research applications of electron microscopy

There are several important diagnostic applications of EM for assisting in the diagnosis of a variety of platelet disorders, particularly for granule defects [15]. Further, there have been recent increases in interest in utilizing EM for platelet research, in particular with the advent of cryo-ET paired with quantitative image analysis of platelet ultrastructural features.

The diagnosis of several granule defects is assisted or can even be made using EM. For example, in Gray Platelet Syndrome, which consists of a rare, heterogenous group of inherited platelet disorders, "gray" and enlarged platelets are observed by EM after Wright staining [59, 60]. More specifically, EM is able to reliably demonstrate the absence of alpha granules and presence of empty vacuoles (which are the immature alpha granule precursors), a key diagnostic criterion [15]. Furthermore, cryogenic electron microscopy studies have revealed different alpha granule subtypes, some of which are in fact tubular in nature and lack vWF [7]. Paris Trousseau syndrome is a rare platelet bleeding disorder also present in nearly all patients with Jacobsen syndrome (chromosomal disorder with multiple anomalies), and the diagnosis can be confirmed by detecting the presence of giant alpha granules by EM with immune-gold labeling [16]. Similarly, for Hermansky-Pudlak syndrome,

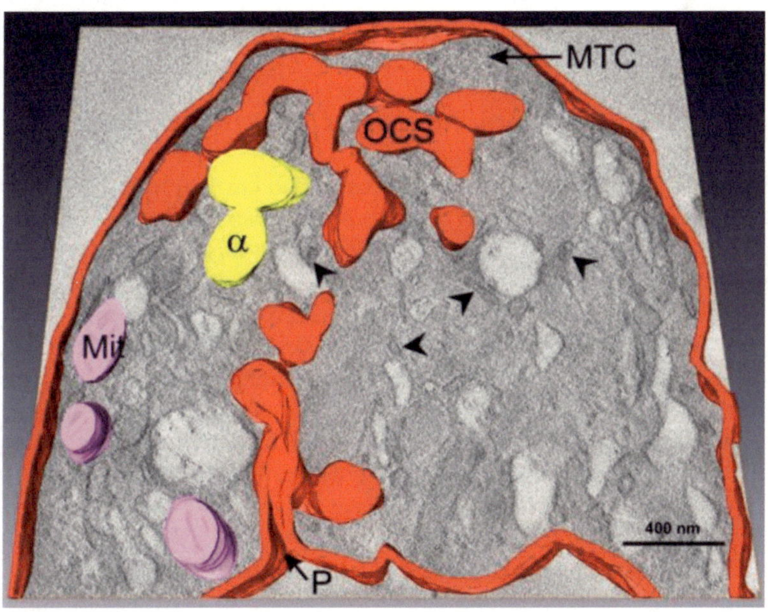

Figure 5.
Electron tomographic 3D model of the open canalicular system (OCS). The OCS is shown with a pore (P) connecting it with the surrounding milieu. The DTS is indicated by arrowheads. One virtual slice is shown in background. Abbreviations: *MTC, microtubular coil DTS dense tubular system; α, α-granules; Mit, mitochondria; OCS, open canalicular system; P, pores of the OCS. Figure and caption reproduced from Neumüller et al. [51], under open access Creative Commons 3.0 Licensing (https://creativecommons.org/licenses/by-sa/3.0/deed.en).*

the absence of dense granules on EM is criteria for diagnosis [17], while Chediak-Higashi syndrome can be confirmed by observing decreased dense bodies, normal alpha granules, and giant cytoplasmic inclusions [46]. While the clinical use of EM is not widely available, it remains an important tool for the diagnosis of rare, but important platelet disorders described above, as well as for several others [15, 46].

In addition to the current clinical applications of platelet EM for the diagnosis of platelet disorders, emerging research using cryo-ET shows promise for expanding our understanding of the role of platelets in a range of disease states including cancer [61], thrombo-inflammatory conditions, and for monitoring the quality of blood banking strategies for platelets [51, 58]. In a recent investigation using cryo-ET of platelets from patients with ovarian cancer, the authors identified important morphological changes in nine parameters between those with and without malignancy, including shortened microtubules, and increases in both size and number of platelet mitochondria [61]. These differences were then incorporated into a prediction model which accurately identified 20 out of 23 patients with ovarian cancer, demonstrating exciting potential for future diagnostic purposes.

Electron microscopy has also been used to characterize the properties of platelet microparticles, patterns of platelet activation, as well as the structure of platelet plug formation in *in vivo* models [18, 58, 62, 63]. For example, Ponomareva et al used both SEM and TEM to study the generation of microparticles from platelets in response to stimulation from adenosine diphosphate, arachidonic acid, and thrombin, and found that many microparticles were in fact smaller (less than 300 nm) than those typically able to be detected by flow cytometry [18]. Furthermore, they found that microparticles isolated from patients with conditions such as disseminated intravascular coagulation, heparin induced thrombocytopenia, or lupus were larger and displayed pathologic properties [18]. Lastly, in an *in vivo* mouse model of vascular injury, Tomaiuolo et al used a combination of fluorescence microscopy and SEM to detail different levels of platelet activation within a forming thrombus, and the effects of antiplatelet agents on the hemostatic plug's architecture [63].

4.5 Current limitations and future directions in electron microscopy

Despite its clear potential as both a research and clinical tool, EM has important drawbacks. In addition to high costs, it requires specialized equipment and highly trained personnel, and therefore is neither widely available nor easy to automate. Furthermore, from a technical perspective, sample thickness is limited, as resolution dramatically decreases with increasing diameters, though this presents more of a challenge for imaging larger cells than it does for platelets [41]. Further, if an intracellular structure takes on an unexpected shape or localization, it can be very difficult to identify without the aid of immuno-labeling, which is not always feasible. Currently, techniques to label intracellular structures require fluorescent tagging and light microscopy, which can be correlated to electron microscopy images (correlative light and electron microscopy or CLEM). These have been improved upon with the use of super-resolution light microscopies as described above, but still does not reach the resolution of EM (super-resolution correlated light and electron microscopy) [35, 36, 41, 64]. However, emerging techniques using cloneable, genetic, electron dense markers, for direct localization with EM could have important applications if used with EM of platelets [65, 66]. Lastly, expanding the use of EM for diagnostic purposes, not just of rare platelet disorders, but in other conditions which affect platelets such as malignancy may provide important insights into disease processes as well as novel diagnostic approaches [61].

5. Intravital microscopy: imaging thrombus formation and platelet function *in vivo*

5.1 Overview

Intravital microscopy is a well-established research tool that has been increasingly applied to study platelets, real-time, *in vivo*. In this technique, the role of platelets in thrombus formation and other cellular activities such as interactions with leukocytes can be directly observed by the surgical creation of "imaging window" in anesthetized rat or mouse models. The major advantage of this approach is that it allows for observation of platelets in an environment that most closely matches what occurs *in vivo* in humans compared to other assays of platelet function. Imaging cells in living organisms has been practiced since the early twentieth century [67]. However, it was not until recently with the development of multiphoton or non-linear optical microscopy that has allowed for imaging in tissues at deeper depths (up to 1 mm) whereas conventional confocal microscopy can only be used up to depths of 50–60 μm [68, 69]. Lastly, with the advent of advanced computational power and digital image acquisition technology integrating with microscopes, quantitative analysis of fluorescent markers and other image properties can now be performed to more comprehensively evaluate properties of clot formation, and distinctive platelet activation and signaling patterns [9, 11, 70, 71].

5.2 Choice of site and imaging technique

The cremaster muscle and mesenteric veins of mice are the most commonly used imaging windows for intravital microscopy, but other possible sites include carotid artery, brain, skin, bones, or liver [11, 72, 73]. In order for visualization and subsequent quantitative analyses, platelets must be labeled with fluorescent antibodies or genetically modified mouse models which express fluorescent platelet proteins. Both of these approaches provide adequate capacity to visualize platelets *in vivo*, however, at least for some of the genetic approaches in mice, there may be untoward effects on platelet aggregation compared to antibody labeling approaches [74]. While any vascular bed theoretically may be used to study platelet function, the primary limitations include the anatomic challenges of isolating the vessels while minimizing trauma and inflammation, as well as vessel caliber, since it must be sufficiently transparent for light penetration [75]. Pros and cons of each anatomic site are summarized below (**Table 1**).

The essential imaging equipment includes a microscope, an illumination light source for fluorescence excitation, an ablation laser (to generate endothelial injury), an image intensifier, and a camera [75]. Endothelial injury and thrombosis can also be induced by mechanical or chemical means, but laser injury offers the advantage of greater temporal and spatial control than the other methods [75]. There are a number of different microscopes that may be used, depending on the type of imaging required, which is in large part dictated by the depth of penetration required by type of tissue being examined, see Masedunskas et al. [69]. Recommendations for camera characteristics include that they be sensitive enough to detect the lowest expected fluorescent signals and to be able to reset fast enough to allow frame rates of 10 per second [75]. For further details on methods of performing intravital microscopy to study platelet function, see Stalker et al. [9], Herr et al. [11], and Falati et al. [70].

5.3 Applications and examples

Intravital microscopy has several applications to identify underlying mechanisms of disease and new aspects of platelet biology. A few illustrative examples of the value

Site	General approach	Advantages	Limitations
Cremaster Muscle	Muscle is exteriorized through scrotal incision in mice	Quick preparation (5–7 min) Thin, transparent Multiple vessels can be studied simultaneously	Vessels are too small to assess flow by Doppler
Mesentery	Mesentery exteriorized through midline incision	Mesenteric vessels large enough for easy ligation or glass pipette mechanical injury	Older mice (>3–4 weeks) have too much fat around vessels to allow for brightfield microscopy Fewer vessels can be studied simultaneously Greater trauma during sample preparation
Ear	Ear is shaved and trans-illuminated to identify vessels	Simple preparation	Difficult to produce reliable injury models due to variation in vascular anatomy
Carotid Artery	Neck incision with dissection of carotid artery	Vessel large enough for Doppler flow monitoring	Only amenable to fluorescent microscopy, no bright field images can be obtained
Brain	Cranial window on dorsal surface of skull	Good for studying microvasculature	Technically challenging
Liver	Lateral abdominal incision with liver exteriorization	Amenable to conventional confocal microscopy	Technically challenging

**See Merrill-Skoloff et al. [75], Falati et al. [70], and Masedunskas et al. [69] for further descriptions of approaches to each site.*

Table 1.
*Anatomic locations for intravital microscopy, advantages and limitations.**

and versatility of the technique include studies demonstrating mechanisms of microvascular thrombosis, platelet production, and for examining platelet leukocyte interactions [19, 72, 73, 76]. For example, in a murine model of arteriolar thrombosis, Lu et al employed cranial intravital microscopy to study anti-platelet properties of caffeic acid (**Figure 6**) [72]. It can also be used to better characterize platelet interactions with other cell types. Using liver intravital microscopy to examine pathways involved in steatohepatitis, Malehmir et al. showed that platelet colonization depended on Kupffer cells and CD44 binding, and that treatment with anti-platelet therapy reduced infiltration of platelets into liver cells as well as platelet immune interactions and trafficking, ultimately leading to attenuation of liver damage [73]. Lastly, the identification of the lung microvasculature as a site of platelet production is another remarkable discovery made possible by the use of intravital microscopy [19].

5.4 Limitations and future advances

The main limitations of intravital microscopy include the depth of penetration depending on the type of microscope used, technical challenges, heterogeneous features and low transparency of certain tissues, and limitations on the number of fluorescent tags that can be used simultaneously [68, 69, 75]. Future advances may include incorporation of correlative light and electron microscopy (CLEM) to intravital approaches, technical advances in miniaturization of lenses and imaging equipment, and improvements in the array of genetic and molecular markers used to track platelet-cellular and platelet-ultrastructural processes [68].

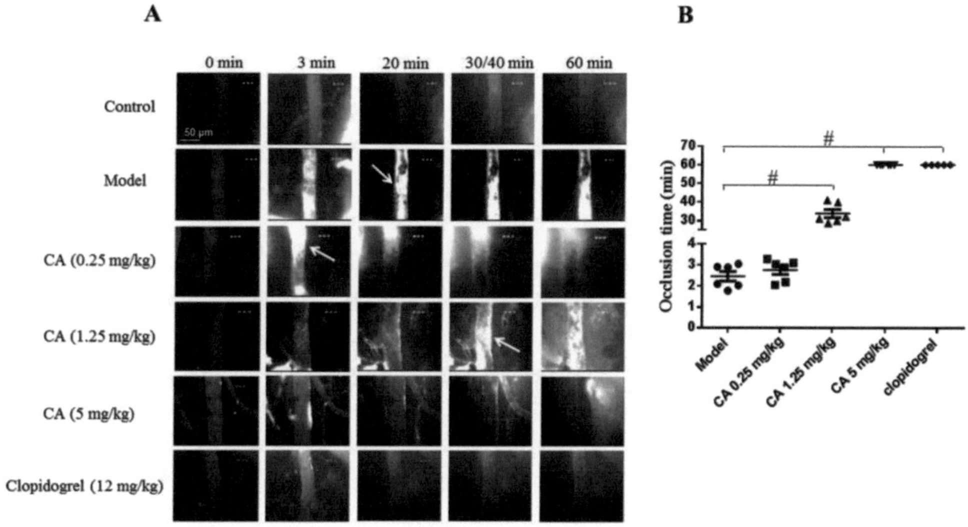

Figure 6.
Cranial intravital microscopy of thrombus formation in a mouse model. (A) Representative images of thrombus formation in mouse cerebral arterioles induced by photochemical injury in different conditions and at different time points. In order to compare the activities of CA and clopidogrel at a similar blood concentration, 5 mg/kg of CA (MW: 180.16) and 12 mg/kg of clopidogrel (MW: 419.90) were continuously infused through the catheterized femoral vein starting from 20 min prior to the vessel wall injury. This resulted in approximately the same blood concentration of CA and clopidogrel in mice in vivo. *Arrows indicate irreversible arteriolar vessel occlusion. (B) Dot plots of arteriolar occlusion time from five to six experiments.* $p < 0.05$ vs *Model group.* Abbreviations: *CA, caffeic acid; MW, molecular weight. Figure and caption reproduced from Lu et al. [76], under the Creative Commons Attribution 4.0 International License (https://creativecommons.org/licenses/by-sa/4.0/deed.en).*

6. Conclusions and future directions

An increasingly versatile armamentarium of imaging modalities is available for research and diagnostic purposes to study platelets [11, 61, 68, 71]. We anticipate these will continue to play instrumental roles in further discoveries in the field of platelet biology, revealing new insights into the dynamics of platelet signaling, ultrastructure, and function. Understanding how to utilize these evolving imaging technologies to best suit the visualization of platelets is essential. Given the inherently highly reactive nature of platelets, newer approaches that cause minimal manipulation and processing offer significant advantages, including intravital microscopy for studying the roles of platelets roles in thrombus formation and cellular interactions [70], and cryogenic electron tomography which preserves platelets in a near native state for ultrastructural analysis [61]. Important advances continue to be made in labeling approaches that can be paired with imaging modalities, including correlative light and electron microscopy, and genetically inducible fluorescent or electron dense markers [34, 65, 66, 77]. In addition to the technical advances in our imaging capabilities, their integration with advanced statistical and computational power will continue to reveal new opportunities to analyze and better understand platelet structure and function.

Conflict of interest

Dr. Kornblith and Dr. Matthay have nothing to disclose.

Funding

Dr. Kornblith is supported by NIH 1K23GM130892-01; Dr. Matthay is supported by The National Center for Advancing Translational Sciences of the NIH award 5TL1TR001871-04.

Other declarations

Human and Animal Rights Informed Consent: This study does not contain any studies with human or animal subjects performed by any of the authors.

Author details

Zachary A. Matthay* and Lucy Zumwinkle Kornblith*
Division of General Surgery, Trauma and Surgical Critical Care, Zuckerberg San Francisco General Hospital, San Francisco, CA, USA

*Address all correspondence to: zachary.matthay@ucsf.edu and lucy.kornblith@ucsf.edu

References

[1] Vieira-de-Abreu A, Campbell RA, Weyrich AS, Zimmerman GA. Platelets: Versatile effector cells in hemostasis, inflammation, and the immune continuum. Seminars in Immunopathology. 2012;**34**(1):5-30

[2] Nachman RL, Rafii S. Platelets, petechiae, and preservation of the vascular wall. The New England Journal of Medicine. 2008;**359**(12):1261-1270

[3] Kerrigan SW, Poole A. Focusing on the role of platelets in immune defence against invading pathogens. Platelets. 2015;**26**(4):285

[4] Hoffman M, Monroe DM 3rd. A cell-based model of hemostasis. Thrombosis and Haemostasis. 2001;**85**(6):958-965

[5] Rondina MT, Weyrich AS, Zimmerman GA. Platelets as cellular effectors of inflammation in vascular diseases. Circulation Research. 2013;**112**(11):1506-1519

[6] Thon JN, Italiano JE. Platelets: Production, morphology and ultrastructure. Handbook of Experimental Pharmacology. 2012;**210**: 3-22

[7] van Nispen tot Pannerden H, de Haas F, Geerts W, Posthuma G, van Dijk S, Heijnen HF. The platelet interior revisited: Electron tomography reveals tubular alpha-granule subtypes. Blood. 2010;**116**(7):1147-1156

[8] Barendrecht AD, Verhoef JJF, Pignatelli S, Pasterkamp G, Heijnen HFG, Maas C. Live-cell imaging of platelet degranulation and secretion under flow. Journal of Visualized Experiments. 2017;**125**:e-55658

[9] Stalker TJ, Traxler EA, Wu J, Wannemacher KM, Cermignano SL, Voronov R, et al. Hierarchical organization in the hemostatic response and its relationship to the platelet-signaling network. Blood. 2013;**121**(10):1875-1885

[10] Cramer EM, Meyer D, le Menn R, Breton-Gorius J. Eccentric localization of von Willebrand factor in an internal structure of platelet alpha-granule resembling that of Weibel-Palade bodies. Blood. 1985;**66**(3):710-713

[11] Herr N, Mauler M, Bode C, Duerschmied D. Intravital microscopy of leukocyte-endothelial and platelet-leukocyte interactions in mesenterial veins in mice. Journal of Visualized Experiments. 2015;**102**:e53077

[12] Kerrigan SW, Clarke N, Loughman A, Meade G, Foster TJ, Cox D. Molecular basis for *Staphylococcus aureus*-mediated platelet aggregate formation under arterial shear in vitro. Arteriosclerosis, Thrombosis, and Vascular Biology. 2008;**28**(2):335-340

[13] White JG. Morphological studies of platelets and platelet reactions. Vox Sanguinis. 1981;**40**(Suppl 1):8-17

[14] White JG. Electron microscopy methods for studying platelet structure and function. Methods in Molecular Biology. 2004;**272**:47-63

[15] Clauser S, Cramer-Borde E. Role of platelet electron microscopy in the diagnosis of platelet disorders. Seminars in Thrombosis and Hemostasis. 2009;**35**(2):213-223

[16] White JG. Use of the electron microscope for diagnosis of platelet disorders. Seminars in Thrombosis and Hemostasis. 1998;**24**(2):163-168

[17] Witkop CJ, Krumwiede M, Sedano H, White JG. Reliability of absent platelet dense bodies as a diagnostic criterion for Hermansky-Pudlak

syndrome. American Journal of Hematology. 1987;**26**(4):305-311

[18] Ponomareva AA, Nevzorova TA, Mordakhanova ER, Andrianova IA, Rauova L, Litvinov RI, et al. Intracellular origin and ultrastructure of platelet-derived microparticles. Journal of Thrombosis and Haemostasis. 2017;**15**(8):1655-1667

[19] Lefrancais E, Ortiz-Munoz G, Caudrillier A, Mallavia B, Liu F, Sayah DM, et al. The lung is a site of platelet biogenesis and a reservoir for haematopoietic progenitors. Nature. 2017;**544**(7648):105-109

[20] Ribatti D, Crivellato E. Giulio Bizzozero and the discovery of platelets. Leukemia Research. 2007;**31**(10):1339-1341

[21] Greinacher A, Pecci A, Kunishima S, Althaus K, Nurden P, Balduini CL, et al. Diagnosis of inherited platelet disorders on a blood smear: A tool to facilitate worldwide diagnosis of platelet disorders. Journal of Thrombosis and Haemostasis. 2017;**15**(7):1511-1521

[22] Balduini CL, Pecci A, Noris P. Diagnosis and management of inherited thrombocytopenias. Seminars in Thrombosis and Hemostasis. 2013;**39**(2):161-171

[23] Sandrock K, Knofler R, Greinacher A, Furll B, Gerisch S, Schuler U, et al. Novel mutation in Bernard-Soulier syndrome. Transfusion Medicine and Hemotherapy. 2010;**37**(5):278-284

[24] Aslan JE, Itakura A, Gertz JM, McCarty OJ. Platelet shape change and spreading. Methods in Molecular Biology. 2012;**788**:91-100

[25] Thomas SG, Calaminus SD, Auger JM, Watson SP, Machesky LM. Studies on the actin-binding protein HS1 in platelets. BMC Cell Biology. 2007;**8**:46

[26] Eshel-Green T, Berny MA, Conley RB, McCarty OJ. Effect of sex difference on platelet adhesion, spreading and aggregate formation under flow. Thrombosis and Haemostasis. 2009;**102**(5):958-965

[27] Calaminus SD, Thomas S, McCarty OJ, Machesky LM, Watson SP. Identification of a novel, actin-rich structure, the actin nodule, in the early stages of platelet spreading. Journal of Thrombosis and Haemostasis. 2008;**6**(11):1944-1952

[28] Calaminus SD, McCarty OJ, Auger JM, Pearce AC, Insall RH, Watson SP, et al. A major role for scar/WAVE-1 downstream of GPVI in platelets. Journal of Thrombosis and Haemostasis. 2007;**5**(3):535-541

[29] Tunstromer K, Faxalv L, Boknas N, Lindahl TL. Quantification of platelet contractile movements during thrombus formation. Thrombosis and Haemostasis. 2018;**118**(9):1600-1611

[30] Sakariassen KS, Aarts PA, de Groot PG, Houdijk WP, Sixma JJ. A perfusion chamber developed to investigate platelet interaction in flowing blood with human vessel wall cells, their extracellular matrix, and purified components. The Journal of Laboratory and Clinical Medicine. 1983;**102**(4):522-535

[31] Claesson K, Lindahl TL, Faxalv L. Counting the platelets: A robust and sensitive quantification method for thrombus formation. Thrombosis and Haemostasis. 2016;**115**(6):1178-1190

[32] Roest M, Reininger A, Zwaginga JJ, King MR, Heemskerk JW. Biorheology subcommittee of the SSCotI. Flow chamber-based assays to measure thrombus formation in vitro: Requirements for standardization. Journal of Thrombosis and Haemostasis. 2011;**9**(11):2322-2324

[33] Van Kruchten R, Cosemans JM, Heemskerk JW. Measurement of whole blood thrombus formation using parallel-plate flow chambers—A practical guide. Platelets. 2012;**23**(3):229-242

[34] Knight AE, Gomez K, Cutler DF. Super-resolution microscopy in the diagnosis of platelet granule disorders. Expert Review of Hematology. 2017;**10**(5):375-381

[35] Westmoreland D, Shaw M, Grimes W, Metcalf DJ, Burden JJ, Gomez K, et al. Super-resolution microscopy as a potential approach to diagnosis of platelet granule disorders. Journal of Thrombosis and Haemostasis. 2016;**14**(4):839-849

[36] Meixner AJ. The Nobel prize in chemistry 2014 for the development of super-resolved fluorescence microscopy. Analytical and Bioanalytical Chemistry. 2015;**407**(7):1797-1800

[37] Schermelleh L, Ferrand A, Huser T, Eggeling C, Sauer M, Biehlmaier O, et al. Super-resolution microscopy demystified. Nature Cell Biology. 2019;**21**(1):72-84

[38] Schermelleh L, Heintzmann R, Leonhardt H. A guide to super-resolution fluorescence microscopy. The Journal of Cell Biology. 2010;**190**(2):165-175

[39] Poulter NS, Pollitt AY, Davies A, Malinova D, Nash GB, Hannon MJ, et al. Platelet actin nodules are podosome-like structures dependent on Wiskott-Aldrich syndrome protein and ARP2/3 complex. Nature Communications. 2015;**6**:7254

[40] Ruska E. Nobel lecture. The development of the electron microscope and of electron microscopy. Bioscience Reports. 1987;**7**(8):607-629

[41] Oikonomou CM, Jensen GJ. Cellular electron cryotomography: Toward structural biology in situ. Annual Review of Biochemistry. 2017;**86**:873-896

[42] White JG. The dense bodies of human platelets. Origin of serotonin storage particles from platelet granules. The American Journal of Pathology. 1968;**53**(5):791-808

[43] White JG. Ultrastructural defects in congenital disorders of platelet function. Annals of the New York Academy of Sciences. 1972;**201**:205-233

[44] White JG. Identification of platelet secretion in the electron microscope. Series Haematologica. 1973;**6**(3):429-459

[45] White JG, Gerrard JM. Ultrastructural features of abnormal blood platelets. A review. The American Journal of Pathology. 1976;**83**(3):589-632

[46] White JG. Platelet microtubules and giant granules in the Chediak-Higashi syndrome. The American Journal of Medical Technology. 1978;**44**(4):273-278

[47] White JG. Ultrastructural studies of the gray platelet syndrome. The American Journal of Pathology. 1979;**95**(2):445-462

[48] White JG. An overview of platelet structural physiology. Scanning Microscopy. 1987;**1**(4):1677-1700

[49] Bessis M, Breton-Gorius J. Microtubules and fibrils in spread-out platelets. Nouvelle Revue Française d'Hématologie. 1965;**5**(4):657-662

[50] Bessis M, Dreyfus B, Breton-Gorius J, Sultan C. Electron microscopic study of 11 cases of refractory anemias with multiple enzymopathies. Nouvelle Revue Française d'Hématologie. 1969;**9**(1):87-104

[51] Neumüller J, Ellinger A, Wagner T. Transmission Electron Microscopy of

Platelets from Apheresis and Buffy-Coat-Derived Platelet Concentrates, the Transmission Electron Microscope—Theory and Applications. Rijeka: IntechOpen; 2015. Available from: https://www.intechopen.com/books/the-transmission-electron-microscope-theory-and-applications/transmission-electron-microscopy-of-platelets-from-apheresis-and-buffy-coat-derived-platelet-concent

[52] Bentfeld-Barker ME, Bainton DF. Identification of primary lysosomes in human megakaryocytes and platelets. Blood. 1982;**59**(3):472-481

[53] Boudreau LH, Duchez AC, Cloutier N, Soulet D, Martin N, Bollinger J, et al. Platelets release mitochondria serving as substrate for bactericidal group IIA-secreted phospholipase A2 to promote inflammation. Blood. 2014;**124**(14):2173-2183

[54] Marcoux G, Magron A, Sut C, Laroche A, Laradi S, Hamzeh-Cognasse H, et al. Platelet-derived extracellular vesicles convey mitochondrial DAMPs in platelet concentrates and their levels are associated with adverse reactions. Transfusion. 2019;**59**(7):2403-2414

[55] Heijnen HF, Oorschot V, Sixma JJ, Slot JW, James DE. Thrombin stimulates glucose transport in human platelets via the translocation of the glucose transporter GLUT-3 from alpha-granules to the cell surface. The Journal of Cell Biology. 1997;**138**(2):323-330

[56] Dubochet J, Adrian M, Chang JJ, Homo JC, Lepault J, McDowall AW, et al. Cryo-electron microscopy of vitrified specimens. Quarterly Reviews of Biophysics. 1988;**21**(2):129-228

[57] Koster AJ, Grimm R, Typke D, Hegerl R, Stoschek A, Walz J, et al. Perspectives of molecular and cellular electron tomography. Journal of Structural Biology. 1997;**120**(3):276-308

[58] Nemeüller J, Meisslitzer-Ruppitsch C, Ellinger A, Pavelka M, Jungbauer C, Renz R, et al. Monitoring of platelet activation in platelet concentrates using transmission electron microscopy. Transfusion Medicine and Hemotherapy. 2013;**40**(2):101-107

[59] Raccuglia G. Gray platelet syndrome. A variety of qualitative platelet disorder. The American Journal of Medicine. 1971;**51**(6):818-828

[60] Drouin A, Favier R, Masse JM, Debili N, Schmitt A, Elbim C, et al. Newly recognized cellular abnormalities in the gray platelet syndrome. Blood. 2001;**98**(5):1382-1391

[61] Wang R, Stone RL, Kaelber JT, Rochat RH, Nick AM, Vijayan KV, et al. Electron cryotomography reveals ultrastructure alterations in platelets from patients with ovarian cancer. Proceedings of the National Academy of Sciences of the United States of America. 2015;**112**(46):14266-14271

[62] Ramtoola Z, Lyons P, Keohane K, Kerrigan SW, Kirby BP, Kelly JG. Investigation of the interaction of biodegradable micro- and nanoparticulate drug delivery systems with platelets. The Journal of Pharmacy and Pharmacology. 2011;**63**(1):26-32

[63] Tomaiuolo M, Matzko CN, Poventud-Fuentes I, Weisel JW, Brass LF, Stalker TJ. Interrelationships between structure and function during the hemostatic response to injury. Proceedings of the National Academy of Sciences of the United States of America. 2019;**116**(6):2243-2252

[64] Chang YW, Chen S, Tocheva EI, Treuner-Lange A, Lobach S, Sogaard-Andersen L, et al. Correlated cryogenic photoactivated localization microscopy and cryo-electron tomography. Nature Methods. 2014;**11**(7):737-739

[65] Wang Q, Mercogliano CP, Lowe J. A ferritin-based label for cellular electron cryotomography. Structure. 2011;**19**(2):147-154

[66] Mercogliano CP, DeRosier DJ. Concatenated metallothionein as a clonable gold label for electron microscopy. Journal of Structural Biology. 2007;**160**(1):70-82

[67] Beck JS, Berg BN. The circulatory pattern in the islands of Langerhans. The American Journal of Pathology. 1931;**7**(1):31-61

[68] Weigert R, Sramkova M, Parente L, Amornphimoltham P, Masedunskas A. Intravital microscopy: A novel tool to study cell biology in living animals. Histochemistry and Cell Biology. 2010;**133**(5):481-491

[69] Masedunskas A, Milberg O, Porat-Shliom N, Sramkova M, Wigand T, Amornphimoltham P, et al. Intravital microscopy: A practical guide on imaging intracellular structures in live animals. BioArchitecture. 2012;**2**(5):143-157

[70] Falati S, Gross P, Merrill-Skoloff G, Furie BC, Furie B. Real-time in vivo imaging of platelets, tissue factor and fibrin during arterial thrombus formation in the mouse. Nature Medicine. 2002;**8**(10):1175-1181

[71] Bellido-Martin L, Chen V, Jasuja R, Furie B, Furie BC. Imaging fibrin formation and platelet and endothelial cell activation in vivo. Thrombosis and Haemostasis. 2011;**105**(5):776-782

[72] Adili R, Holinstat M. Formation and resolution of Pial microvascular thrombosis in a mouse model of thrombotic thrombocytopenic Purpura. Arteriosclerosis, Thrombosis, and Vascular Biology. 2019;**39**(9):1817-1830

[73] Malehmir M, Pfister D, Gallage S, Szydlowska M, Inverso D, Kotsiliti E, et al. Platelet GPIbalpha is a mediator and potential interventional target for NASH and subsequent liver cancer. Nature Medicine. 2019;**25**(4):641-655

[74] Da Q, Derry PJ, Lam FW, Rumbaut RE. Fluorescent labeling of endogenous platelets for intravital microscopy: Effects on platelet function. Microcirculation. 2018;**25**(6):e12457

[75] Merrill-Skoloff G, Atkinson BT, Furie BC, Furie B. Real time in vivo imaging of platelets during thrombus formation. In: Michelson A, editor. Platelets. 3rd ed. Netherlands: Elsevier; 2013. pp. 635-652

[76] Lu Y, Li Q, Liu YY, Sun K, Fan JY, Wang CS, et al. Inhibitory effect of caffeic acid on ADP-induced thrombus formation and platelet activation involves mitogen-activated protein kinases. Scientific Reports. 2015;**5**:13824

[77] Chen D, Uhl CB, Bryant SC, Krumwiede M, Barness RL, Olson MC, et al. Diagnostic laboratory standardization and validation of platelet transmission electron microscopy. Platelets. 2018;**29**(6):574-582

Chapter 2

Molecular Aspects of Pathophysiology of Platelet Receptors

Mrinal K. Poddar and Soumyabrata Banerjee

Abstract

Receptor is a dynamic instrumental surface protein that helps to interact with specific molecules to respond accordingly. Platelet is the smallest in size among the blood components, but it plays many pivotal roles to maintain hemostasis involving its surface receptors. It (platelet) has cell adhesion receptors (e.g., integrins and glycoproteins), leucine-rich repeats receptors (e.g., TLRs, glycoprotein complex, and MMPs), selectins (e.g., CLEC, P-selectin, and CD), tetraspanins (e.g., CD and LAMP), transmembrane receptors (e.g., purinergic—P2Y and P2X1), prostaglandin receptors (e.g., TxA2, PGH2, and PGI2), immunoglobulin superfamily receptors (e.g., FcRγ and FcεR), etc. on its surface. The platelet receptors (e.g., glycoproteins, protease-activated receptors, and GPCRs) during platelet activation are over expressed and their granule contents are secreted (including neurotransmitters, cytokines, and chemokines) into circulation, which are found to be correlated with different physiological conditions. Interestingly, platelets promote metastasis through circulation protecting from cytolysis and endogenous immune surveillance involving several platelets receptors. The updated knowledge about different types of platelet receptors in all probable aspects, including their inter- and intra-signaling mechanisms, are discussed with respect to not only its (platelets) receptor type but also under different pathophysiological conditions.

Keywords: platelet receptors, thrombus, cancer, aging, cardiovascular disease, viral infection

1. Introduction

Platelets including other blood components were first time drawn by George Gulliver in 1841 (though it was not named platelet at that time) when he was working with a newly made compound microscope with twin lens [1]. Giulio Bizzozero at the end of the nineteenth century [2] first time coined the term and identified as platelets. Blood platelet in its form is found in mammals, but in birds and amphibians, it is present in circulation as intact mononuclear thrombocytes [3]. Platelets are not true cells, and it is classified as cell fragments (from megakaryocyte by the megakaryocytopoiesis) having no nucleus inside. Platelets circulate in the bloodstream and remain alive for 7–10 days. They (platelets) principally survey the inner lining of blood vessels. If they detect any breaches, they seal them in the vasculature by the creation of thrombi [4]. Platelets generally remain in inactive state

Sl. no.	Classes of receptors	Types of receptors	Family	Ligands	Involvement	References
1.	GP	GPIb-IX-V complex	Type I membrane spanning GP and leucine-rich repeat	vWF, thrombin, P-selectin, $\alpha M\beta 2$, and Mac-1	Initiation of platelet recruitment	[5–22]
		GPVI	Immunoglobulin (Ig)	Collagen and laminin	Platelet aggregation	[23–26]
		GPVI-FcRγ	Transmembrane	Collagen and laminin	Adhesion with collagen	[7, 27, 28]
		Integrins ($\alpha IIb\beta 3$, $\alpha v\beta 3$, $\alpha 2\beta 1$, $\alpha 5\beta 1$, and $\alpha 6\beta 1$)	Transmembrane	Fibrinogen or vWF, vitronectin, collagen, fibronectin, and laminin, respectively	Platelet aggregations	[7, 29–35]
2.	C-type lectin receptor	CLEC-2	Type II membrane protein C-type lectin receptor	Podoplanin and rhodocytin	Platelet aggregation	[36, 37]
3.	Thromboxane	TP (α and β)	Transmembrane and GPCR	TXA_2	Platelet aggregation	[38–40]
4.	Prostaglandin (PG)	PGE2 and PGI2	GPCR	PG	Aggregation as well as inhibition of aggregation	[41–45]
5.	Thrombin	PAR-1 and -4	Transmembrane	Thrombin	Adhesion, spreading, and secretion	[19, 46]
6.	Ephrin kinase	EphA and EphB	Tyrosine kinase	Ephrins	Inhibition in aggregation	[47, 48]
7.	Purinergic	P2Y1 and P2Y12	Transmembrane and GPCR	ADP	Amplification of aggregation	[49–51]
		P2X1		ATP		
8.	TAM	TAM	TAM tyrosine kinase	Gas6	Stabilization of aggregation	[7, 52, 53]
9.	Tetraspanin	CD151	Tetraspanin	Fibrinogen	Aggregation, stabilization of aggregates	[7, 54–65]
		TSSC6		Not known		
		TLT-1		Not known		
		CD36		vWF, oxPL, TSP1, and oxLDL		
		PEAR1		FcεR1α		
10.	P-selectin	P-selectin	Selectin	PSGL-1	Clot formation with leukocytes	[7, 65]

Sl. no.	Classes of receptors	Types of receptors	Family	Ligands	Involvement	References
11.	ITIM	PECAM-1	Transmembrane and Ig	PECAM	Inhibition in thrombus formation	[66–68]
		G6b-B	Transmembrane and Ig	HS	Platelet production and activation	
		VPAC1	Transmembrane and Ig	PACAP	Inhibition in platelet activation	
12.	TLR	TLR-1, -2, -3, and -4	Lipoprotein	Peptidoglycan and pathogen	Inflammatory response	[69–72]
13.	Serotonin (5-HT)	5-HT 2A and 5-HT3	Transmembrane	Serotonin	Thrombus formation, vasodilation	[72–76]
14.	Leucine-rich receptors	GPIb-IX-V complex	Transmembrane and Ig	vWF, thrombin, P-selectin, αMβ2, and Mac-1	Initiation of platelet recruitment	[5–22, 69–72]
		TLRs	Lipoprotein	Peptidoglycan and pathogen	Inflammatory response	
15.	Complement receptors	CR2, CR3, CR4, C3aR, C5aR, gC1qR, and cC1qR	Complement	Pathogens	Antimicrobial host defense	[69, 77]
16.	DC-SIGN	DC-SIGN	Non-integrin	Viral pathogens	Binding platelet with viral pathogens	[69, 78]

GP: glycoprotein; vWF: von Willebrand factor; TP: TXA_2/PGH2 receptor; CLEC-2: C-type lectin-like type II transmembrane receptor; GPCR: G protein-coupled receptor; TX: thromboxane; PAR: protease-activated receptors; Eph: ephrin; TAM: Tyro 3, Axl, and Mer; TSSC6: tumor suppressing STF cDNA 6; Gas: growth arrest specific; TLT: TREM-like transcript-1; PEAR: platelet endothelial aggregation receptor-1; PECAM-1: platelet endothelial cell adhesion molecule; VPAC1: vasoactive intestinal peptide/pituitary adenylate cyclase-activating peptide receptor 1; PACAP: pituitary adenylate cyclase-activating polypeptide; TLR: Toll-like receptor; 5-HT: 5-hydroxytryptamine, serotonin; DC-SIGN: dendritic cell-specific ICAM3-grabbing non-integrin.

Table 1.
Classes of platelet receptors and their involvement in the maintenance of physiological hemostasis.

and get activated only when a blood vessel is damaged. But hemostasis or blood coagulation is not the sole function of platelets; rather, it is employed in several multifunctional attributes monitoring the homeostasis of the body. The advancement of understanding of platelet pathology has leaded scientists to go through the receptors for its instrumental role in different pathological conditions. There are 16 different classes of platelet receptors. Each class of some of these receptors has subtypes (**Table 1**). Platelet receptors (**Figure 1**), the surface proteins, are at the forefront of the recent research, and major advances have been made in understanding their molecular functions as well as their downstream signaling pathways. The experimental strategies with pharmacological inhibition and knocking out

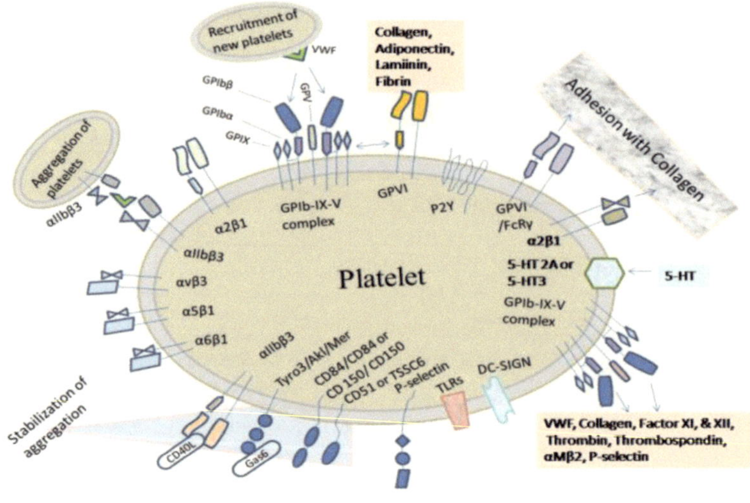

Figure 1.
Different predominant platelet receptors and their physiological role. DC-SIGN: dendritic cell-specific ICAM3-grabbing non-integrin; 5-HT: 5-hydroxytryptamine, serotonin. Other receptor details are same as described in the legends of ***Table 1****.*

of nearly all known receptors and many signaling molecules have helped to reveal new mechanisms for how the thrombotic and hemorrhagic propensity of platelets is controlled in health and disease [79]. The firm adhesion of platelets to the injury site forms first a monolayer. This monolayer acts like a reactive site for other free moving platelets to replenish further. The accumulation of platelets is found to depend on vWF and fibrinogen that act like cross-linking adjacent to the platelets and promotes a stable aggregated platelets or clot [80, 81]. This process is essential for the formation of primary hemostatic clot and also for the development of pathological thrombi at site ruptured [82]. Further activation in developing thrombus converts platelets to a procoagulant phenotype, from a proaggregatory phenotype, which enables the assembly of the coagulation reaction complexes (the tenase and prothrombinase complex) on the platelet surface, necessary for thrombin and fibrin generation. These fibrins form a fibrin network and trap the RBCs and convert the white thrombus to the red clot [83]. This supports secondary hemostasis or red clot [83]. However, physiological thrombus formations or mural thrombi are found to be tightly regulated to avoid the excessive platelet accumulation at the site of injury. The vascular obstructions that are the principal pathological processes, to some extent, cause heart attacks and/or ischemic stroke as well [84]. During initiation, progression, and termination of this physiological thrombus formation, several receptors of platelet get involved as soon as the damage on the blood vessel wall happens. There are different surface receptors on the platelet which are involved not only in blood coagulation process but also in other immunogenic and/or pathogenic responses, even in neurodegeneration. In this chapter, the molecular aspects of different classes of platelet receptors and their contribution in different pathophysiological conditions are discussed.

2. Glycoprotein (GP) receptors

On the membrane of platelet, a class of glycoprotein (GP) receptors is present which plays a key role in hemostasis. In fact, the blood vessel wall damage initiates the event of interaction of GP receptors with the extracellular matrix (**Figure 1**).

The GP receptors include:

- GPIb-IX-V complex,
- GPVI,
- GPVI-FcRγ, and
- integrins.

2.1 GPIb-IX-V complex

The GPIb-IX-V complex consists of glycoproteins (GP) Ibα, Ibβ, IX, and V, which are known as GPIbα (~135 kDa), GPIbβ (~25 kDa), GPIX (~20 kDa), and GPV (~85 kDa), respectively. The GPIbα is linked with a disulfide bond to GPIbβ and non-covalently associated with GPIX and GPV in a ratio of 2:2:2:1 [5, 6]. GPIbα, GPIbβ, and IX are present at almost 25,000 copies per platelet, whereas GPV at 12,500 copies which can be calculated easily by their abundance ratio in a complex. GPIbα, GPIbβ, and GPIX are found to be closely associated and the arrangements of these subunits are essential for an efficient bioavailability of GPIb to its ligand. Lack of any of the single subunits is significantly able to decrease the surface expression of the whole complex (GPIb), whereas GPV is more loosely associated with the complex and its absence does not interfere with the expression of GPIb as well as its interaction with its ligand, vWF. GPV has been found to be active only during the binding of thrombin to GPIbα subunit. It (complex) is found to be assembled in the megakaryocyte in the bone marrow (the origin site of platelet) as a functional unit [7]. These are type I membrane spanning GP and are members of the leucine-rich repeat family, with one or more approximately 24-residue leucine-rich repeats, with their N- and C-terminal disulfide-looped flanking sequences, in their extracellular domains [5]. The four subunits are encoded by genes mapping to chromosomes 17p12 (GPIBA), 22q11.2 (GPIBB), 3q29 (GP5), and 3q21 (GP9) [7]. GPIbα, the major ligand-binding subunit, has a globular N-terminal ligand-binding domain elevated from the cell surface by a sialomucin core [5]. This N terminal (282 residues) contains the leucine-rich repeats (LRR), the flanking sequences, and an anionic sequence (residues 269–282) with the sulfation (Tyr 276–279, except 277). The 1–282 sequence contains overlapping, but distinct binding sites for (a) vWF, (b) leukocyte integrin αMβ2, and (c) P-selectin, a granule-membrane receptor expressed on the surface of activated platelets or activated endothelial cells [5, 8, 9]. The adhesion of vWF to GPIbα under shear stress involves electrostatic interactions between a negative patch (centered on residues 59–128) within the leucine-rich repeats of GPIbα and a complementary positive patch (centered on residues 496–709) on the vWF-A1 domain [5, 10, 11]. The residues of N- and C-terminal to this region directly contact vWF in a co-crystal structure of a GPIbα fragment which lacks N-linked glycosylation sites at Asn21 and Asn159 [12]. The point mutations in either GPIbα (Met239/Val) or vWF-A1 (Arg543/Gln) alter their conformation producing subtle differences in their ligation as gain of function [12, 13]. More preciously, in the case of vWF-A1, the mutation is 415A from a GPIbα contacting sequence, emphasizing the sensitivity of interaction to the conformational regulations [12]. The gain-of-function mutation at Arg543/Gln and Met239/Val enhances affinity for GPIbα by virtue of almost 6-fold. The Met239/Val gain-of-function mutation of GPIbα is within a b-hairpin structure in the C-terminal flank, which undergoes significant structural alteration in a complex with vWF-A1. This Met239/Val gain-of-function mutation stabilizes the β-hairpin and increases the affinity of vWF binding [12]. These conformational changes of native GPIbα and vWF with isolated ligand and receptor fragments might regulate the affinity of the adhesive interaction and result in an on-rate/off-rate for platelet adhesion in either rolling or firm adhesion in flowing blood, though the *in vivo* study is still under progress to understand. Thrombin binding to GPIbα subunit presents the thrombin to its receptor, the protease activated receptor 1 (PAR-1). GPV acts as a negative modulator (as GPV cleaves by thrombin) of thrombin-induced platelet activation, which unmasks GPIb-IX complex and facilitates the binding of thrombin to GPIbα [13]. GPIb interacts with vascular P-selectin, pointing to its function in inflammatory platelet pathways [9]. Depending on the orientation of the active site on GPIbα-bound thrombin, thrombin-mediated proteolysis of platelet surface substrates (including GPV or PAR-1) may be regulated by interaction with GPIbα. GPIbα is a cofactor for PAR-1 activation [9]. In turn, thrombin can regulate GPIb-IX-V signaling directly by engagement of GPIbα under conditions where GPV is absent [9–15]. GPIbα-associated thrombin also regulates Factor XI; the latter also binds to GPIbα [16]. The extracellular domain comprises the binding site for vWF, P-selectin, and Mac-1 (**Table 1**). In addition to its dynamic role in platelet recruitment onto vWF, the GPIb membrane complex functions as a receptor for coagulation factor XII [17], XI [18], thrombin [19, 20], and HK (high-molecular-weight kininogen) [21]. Hence, GPIb is a receptor linking primary and secondary hemostasis [7]. The Src and Lyn of the Src family kinases (SFKs) are associated with the GPIbα subunit to initiate the inside-out signaling. Binding of vWF to the extracellular region of GPIbα subunit induces SFK activation and phosphorylation of downstream substrates. These include immunoreceptor tyrosine-based activation motif (ITAM)-containing FcR γ-chain (FcRγ) and FcγRIIA. Both of them are found to be acted as high-affinity docking sites for the tandem SH2 domain-containing protein-tyrosine kinase Syk. Fc receptors (FcRγ chain and FcγRIIA) may also be associated with GPIb-IX-V on platelets and contribute to GPIb-IX-V-dependent platelet activation. Intracellular signals emanates from GPIb-IX-V [5, 6, 11, 22]. It promotes elevation of cytosolic Ca^{2+}, cytoskeletal changes, secretion of agonists such as ADP (that activates G protein-coupled receptors, P2Y1 and P2Y12), and activation of the integrin αIIbβ3 that binds vWF or fibrinogen and mediates platelet aggregation. In vivo, plasma vWF is not recognized by GPIb-IX-V on resting platelets, thus preventing platelet aggregation in the normal circulation, with the interaction being triggered when sub-endothelial vWF is exposed following injury. GPIb-IX-V complex, in fact, plays a leading role in the elimination of high-stress injury. However, GPIb-IX-V can be induced to bind vWF in plasma at high pathological shear stress (e.g., coronary artery is blocked by atherosclerotic plaque) [7].

2.2 GPVI

Platelet GPVI is a member of the T cell receptor family and immunoglobulin superfamily. It has two extracellular Ig domains: (a) mucin-like domain, a transmembrane domain, and (b) cytoplasmic tail (**Table 1**). These are expressed constitutively on platelets and are engaged during the exposure of collagen in the subendothelial matrix after disruption of the endothelium *in vivo*. Ligands for GPVI including collagen, cross-linked collagen-related peptides, and the snake toxin, convulxin [85] may bind distinct sites of the Ig domains [23, 24]. GPVI can signal in response to collagen or other ligands by FcRγ receptor-dependent or receptor-independent pathways. The dependent pathway involves activation of Syk, and the independent pathway is regulated by calmodulin (via Ca^{2+} signaling pathways) or Fyn/Lyn kinase [7].

The interaction of calmodulin with GPVI produces a soluble GPVI fragment [25]. Engagement of GPVI, the platelet integrins along with the collagen-binding receptors α2β1 and αIIβ3 are found to be up regulated in the process of platelet aggregation [26]. The interaction between GPVI and collagen may be involved in the initiation of thrombus formation either at low or high shear rates. Further in the latter case, this thrombus formation possibly supports platelet aggregation, mediated by the GPIb-IX-V complex and the vWF [26]. These signaling pathways result in secretion of agonists such as ADP and inside-out activation of platelet integrins, primarily αIIbβ3 that binds fibrinogen or vWF and mediates platelet aggregation [26].

2.3 GPVI-FcRγ

Glycoprotein VI (GPVI), a transmembrane protein (63 kDa) consisting of two Ig-like domains in the extracellular region in platelets (**Table 1**). It connects to a highly glycosylated linker (a transmembrane domain) and a cytoplasmic tail. GPVI is expressed exclusively in platelets and megakaryocytes. In platelets, it expressed with around 3700 copies per platelet. The transmembrane adapter protein FcRγ is found to be associated with this. Surface expression of GPVI depends on FcRγ stabilization. This stabilization occurs through a salt bridge between the GPVI transmembrane domain residue Arg272 and FcRγ (Asp residues). The FcRγ is a covalent-linked homodimer containing one copy of an ITAM in each chain. It has two YxxL sequences separated by seven amino acids [27]. Phosphorylation of the ITAM motif by two Src kinases (Fyn and Lyn) associated with GPVI initiates platelet signaling, leading to potent platelet activation. SFKs for signal transmission are either associated with or in close proximity to their cytoplasmic tails. SFKs' downstream effectors (adaptors, enzymes, and cytoskeletal proteins) collectively coordinate cytoskeletal remodeling, degranulation, membrane flipping, and integrin activation, and hence platelet activation. It has also been found that the SFKs act via the GPCRs. The G_q coupled with PAR-1 and PAR-4, and the G_i coupled with ADP receptor, $P2Y_{12}$ which synergizes with the primary activation, signal to maximally activate platelets. GPVI is expressed in platelets as a mixture of monomers and dimers, with a stoichiometry of one GPVI to each FcRγ-chain covalent dimer. The dimeric GPVI forms a unique conformation with higher affinity for collagen to mediate activation at the physiological concentrations of collagen, but not the monomeric GPVI due to its low affinity toward collagen [28]. F(αβ)2 fragments of antibodies to this structure induce platelet activation, while Fab fragments block activation by collagen, indicating a minimal signaling model in which activation is achieved through cross-linking of two GPVI dimmers [7]. Ligand-mediated clustering of different platelet receptors (GPIb-IX-V complex, GPVI-FcR γ, integrins α2β1 and αIIbβ3, hemi-ITAM-containing podoplanin receptor CLEC-2, ITAM-containing low-affinity immunoglobulin receptor FcγRIIA) trigger transmission of primary activation signals through the phosphorylation of downstream tyrosine residues in proteins. These receptors depend on the family of protein-tyrosine kinases, known as Src, but not on the intrinsic kinase activity.

2.4 Platelet integrins

Integrins play an important role in the cell metabolism of every cell, including platelets. The integrin ligands of platelets are partly extracellular matrix bound and insoluble, and partly soluble. Platelet adhesion and aggregation are mediated by the heterodimeric receptors, the β1 and β3 of integrins family. Integrins are expressed in a low-affinity state in the resting platelets, but to bind to their ligands efficiently in response to the cellular activation they (integrins) are found to be shifted to a high-affinity state [29]. The integrin receptor for collagen on platelets, the α2β1, bears important roles at their disposal. A major downstream consequence of engagement of primary platelet adhesive receptors such as GPIb-IX-V and GPVI does rapidly activate platelet integrins.

Platelet integrins (**Table 1** and **Figure 1**) are:

- αIIbβ3 (binds fibrinogen or vWF),
- αvβ3 (binds vitronectin),
- α2β1 (binds collagen),
- α5β1 (binds fibronectin), and
- α6β1 (binds laminin).

The Ca^{2+}-dependent "inside-out" activation of αIIbβ3 to bind vWF is critically involved in stable thrombus formation at high shear stress. In the absence of αIIbβ3, activation of other integrins (e.g., α5β1-binding fibronectin) can at least partly compensate for this role [7]. Recent structural studies provide insight into the molecular mechanism of integrin activation [30]. Simultaneous with these conformational changes, altered attachment of the cytoplasmic domain with cytoskeletal components (e.g., αIIbβ3 binding to talin) facilitates integrin-dependent cell adhesion, signaling, and contraction.

2.4.1 αIIbβ3

The αIIbβ3 is the most abundant and dominant surface-expressed integrin in platelets (**Figure 1**). It may vary from 40,000 to 80,000 copies per platelet. Additional pool of this receptor can be recruited from internal membranes upon agonist-induced platelet activation. It is also the major functional integrin receptor on the platelet surface. The mature αIIb and β3 subunits are 148 and 95-kDa proteins, respectively. The αIIbβ3 binds with several RGD (Arg-Gly-Asp) motif containing ligands including fibrinogen, fibrin, vWF, vitronectin, fibronectin, and thrombospondin. During ligand recognition on platelet surface through RGD tract, fibrinogen (the major platelet αIIbβ3 ligand) promotes cell attachment by initiating αIIbβ3 clustering and recruitment of intracellular proteins. The RGD motif subsequently acts as a molecular switch on the β3 subunit to induce a conformational change necessary for full cell spreading [31]. This integrin mediates platelet aggregation via the SFK signaling pathways through the binding of plasma fibrinogen. It serves as the principal receptor for platelet adhesion *in vivo* [7, 26] with the inside-out signaling. The shifting from a low- to a high-affinity state of integrin αIIbβ3 is being considered the "final common pathway" of platelet activation. It (shifting) is essential for platelet αIIbβ3 to interact with the fibrinogen during platelet adhesion. Needless to mention a well-known fact that fibrinogen itself is a ligand with two receptor interaction sites: (a) enabling interaction with separate platelets and (b) constituting the basis of platelet aggregation. At this stage of transmitting signals, the Src is found to be the most abundant SFK in human platelets and is essential for propagation of signals from the activated αIIbβ3 integrins.

2.4.2 αvβ3

The αvβ3 receptors express widely in endothelial cells, osteoblasts, smooth muscle cells, and leukocytes, and throughout the vascular bed. It is present in only a few hundred copies per platelet. A distinguished difference lies between αvβ3 and αIIbβ3 in platelets. The αvβ3 can bind several RGD containing ligands, including osteopontin (a class of protein which involves diverse physiological functions) and adenovirus penton base (a major capsid protein of human adenovirus), but vitronectin (a glycoprotein, binds to integrin αvβ3, and thus promotes cell adhesion and spreading) is a preferred ligand for αvβ3 [7]. High-affinity αvβ3 can be induced by agonists (adenosine diphosphate, ADP) and by direct integrin modulators (dithiothreitol and $MnCl_2$). This high affinity and activated αvβ3 on platelets can bind osteopontin in atherosclerotic plaques and in the wall of only injured arteries [27, 32].

2.4.3 α5β1

It has been found to be the crucial one to involve the resting platelet to bind with the fibronectin. The α5β1 is the principal platelet receptor which supports resting platelet to adhere with the matrix fibronectin through its RGD sequence in static conditions. However, this interaction is unable to promote calcium oscillation, tyrosine phosphorylation, and/or lamellipodia formation. The interaction of α5β1 with fibronectin is sensitive to shear stress (the tangential force of the flowing blood on the endothelial surface of the blood vessel), and it has been found that it loses its avidity quickly with the increase of the shear stress. Therefore, α5β1 may have the limited role to initiate the interaction of resting platelets with the fibronectin matrix. Especially during injuries in the larger blood vessels, where shear forces are low, it promotes the engagement of other subsequent integrins and also receptors to amplify platelet-induced responses [7].

2.4.4 α6β1

The α6β1, a principal laminin receptor of platelets, does not require any platelet activation in order to bind laminin (a fibrous protein present in the basal lamina of the epithelia, influencing cell differentiation and migration) to promote adhesion. Some cations (Mn^{2+}, Co^{2+}, and Mg^{2+}) support adhesion, while few others (Ca^{2+}, Zn^{2+}, and Cu^{2+}) do not help. Binding of platelets to laminin through α6β1 does not induce platelet aggregation but adherent of platelets to laminin triggers signaling pathways. These signaling pathways induce filopodia formation with PI3K and cdc42 activities in higher rate than in platelets which activates through αIIbβ3 involvement. Laminin has been known for many years to support adhesion of platelets through integrin α6β1, but it has the ability to activate GPVI. The interaction of laminin with GPVI depends on the initial interaction with integrin α6β1. This is just in contrast to the event of collagen interaction which initiates platelet activation through GPVI. This difference between these interactions of these two matrix proteins may reflect the lower affinity (approximately 10-fold) of laminin for GPVI or the presence of a subpopulation of constitutively active α6β1. The weak nature of the GPVI activation (by laminin) argues against a significant role in the prevention of major bleeds. It is suited ideally to facilitate vessel repair after minor damage without the risk of forming occlusive thrombi in the blood vessels [33].

Collagens are not only the most abundant proteins (20–40% of total proteins in the aorta) in the subendothelial extracellular matrix but also it is essential in platelet adherence and platelet plug formation to provide a mechanical strength to the blood vessel wall. There are nine types of collagen resides in the vasculature. Among those, only fibrillar collagens of types I, III, V, and VI and nonfibrillar collagens of types IV and VIII are thrombogenic. Although platelets have various types of receptors for collagen, the receptors (e.g., GPVI, α2β1, p65, p47, TIIICBP, GPIV, the integrin α2β1, and GPVI) are considered as its (platelets) major receptors for binding to collagens and activation of platelets [34].

2.4.5 α2β1

The integrin α2β1 (or VLA2, CD49b/CD29, GPIa/IIa) is a collagen receptor composed of a α2 chain (150 kDa) and a β1 chain (130 kDa). This α2β1 has the expression profile ranging from 900 to 4000 copies/platelet and is expressed per platelet at about 2000 copies. The α2, inserted with 200-residue sequence, is the only platelet subunit with I domain. The crystal structure depicts that this domain comprises (a) seven helices surrounding a core of five parallel β-strands, (b) a short anti-parallel β-strand, and (c) a C-terminal helix [27]. The α2β1 integrin of this I domain binds with collagen and preferably Mg^{2+}/Mn^{2+}, in presence of the metal ion coordinating residues (D151, T221, and D254) in this site. Several recognition sequences (GFOGER, GLOGER, GASGER, GROGER, and GLOGEN) have been identified in collagens I and III. Their recognition profile in this recognition sequences have hierarchy and affinity differently [35]. This interaction is dependent on Mg^{2+} and the GER sequence. α2β1 recognizes these sequences in the resting state, but platelet activation by classical agonists via intracellular signal transduction pathways activates α2β1 via a structural rearrangement of the α2β1 domains [36], causing them to upregulate their affinity for their preferred ligand sequences, stereochemically positioned at regular positions on bundled collagen fibrils [35].

3. C-type lectin-like receptor

Another type of adhesion receptor is C-type lectin (CLEC), a type of carbohydrate-binding protein domain. It requires calcium (that is why “C” comes) for binding. Proteins that contain C-type lectin domains have a diverse range of functions including cell-cell adhesion, immune response (to pathogens), and apoptosis. This kind of receptor is present on platelet to help in adhesion as another adhesion receptor.

3.1 CLEC-2

CLEC-2 is highly expressed in megakaryocytes and platelets, and at low level in mouse neutrophils. The CLEC-2 gene (on chromosome 12) codes for a type II membrane protein C-type lectin receptor family with an extracellular carbohydrate-like recognition domain (CRD-like). It has a cytoplasmic tail of 31 amino acids that contain a single conserved YxxL sequence (known as a hem-ITAM). CLEC-2 was discovered from the snake (the Malayan pit viper, *Calloselasma rhodostoma*) venom rhodocytin (**Table 1**), known as aggretin. Previously it (rhodocytin) was considered to be a platelet activator through the action of α2β1 and GPIbα. This concept was based on the ability of high concentrations of antibodies to block the activation. Now it has been found that rhodocytin does not bind to the recombinant α2β1, as it has the ability to activate platelets even in the deficiency of integrin α2β1, GPIbα, and GPVI. Thus, it is a proven fact now that this rhodocytin appears to activate platelets through a novel receptor, CLEC-2. The CLEC-2 antibody establishes CLEC-2 as a novel platelet activation receptor as it has the ability to induce the potent activation to human platelets [37]. Experimenting with ligand and anti-CLEC-2 antibody, it has been found that Syk mediates phosphorylation of CLEC-2 with Src family kinases plays a critical role in further downstream signaling [7]. The rhodocytin-induced platelet aggregation also previously known to depend on secondary mediators (e.g., thromboxane A_2, TXA_2, and ADP) acted as agonists for GPCRs on platelets. Recently, it has been found that CLEC-2-induced Syk and PLCγ2 phosphorylation potentiates by the TxA2 by playing a critical role in the most proximal event of

CLEC-2 signaling (i.e., CLEC-2 receptor tyrosine phosphorylation). In addition, it may be mentioned that the ADP receptors and protease-activated receptors can also potentiate CLEC-2 signaling during the process of thrombosis formation. The PLCβ-PKCα pathway possibly is regulating the activation of SFKs (mentioned earlier), which are crucial for initiation of CLEC-2 signaling by the G_q-coupled receptors, not other G-proteins.

Amplification of platelet activation by TXA_2 synthesis and binding to the TXA_2/prostaglandin H_2 (TP) receptor are the process of the aspirin-sensitive platelet activation (**Figure 1**). On the other hand, prostacyclin (PGI2) and PGD2 are known to inhibit platelet aggregation, whereas PGE2 potentiates or inhibits platelet response in a dose-dependent manner [7]. In this context, it may be mentioned that bioactive lipid mediators, prostanoids, formed from arachidonic acid by the cyclooxygenase enzyme is known to liberate from the cell membrane. They are involved in numerous physiological activities, including platelet aggregation, local inflammatory response, leucocyte-endothelial cell adhesion, and vasorelaxation as well as vasoconstriction. So, the thromboxane as well as PGs are other types of platelet adhesion receptors to discuss with.

4. Thromboxane receptor

TXA_2 is produced from its precursor arachidonic acid through the cyclooxygenase pathway [38]. The TP receptor or TXA_2/PGH2 receptor (57 kDa) is a membrane-bound seven transmembrane spanning G protein-coupled (including Gq and G12/13) receptor and widely present in the cardiovascular system (**Table 1**). Human TP receptors (TPα and TPβ) are encoded by the same gene, but different from each other as they produce from alternative splicing and have different C-terminal intracytoplasmic regions. Both TPα and TPβ mRNAs are found in platelets, but in endothelial cells, only TPβ has been found to be expressed. TP receptors are also expressed in other cell types related to atherothrombosis (smooth muscle cells, macrophages, and monocytes) [39, 40].

5. Prostaglandin (PG) receptors

The PG receptors, especially PGE2 and PGI2, bear a pivotal responsibility in platelet aggregation phenomena to maintain hemostasis, as mentioned earlier.

5.1 Prostaglandin E2 receptors (PGE2)

The biosynthesis of PGE2 is enhanced by inflammatory mediators in vascular smooth muscle cells and macrophages (**Table 1**). PGE2 shows a biphasic, concentration-dependent effect on platelet aggregation as (a) high concentrations of PG inhibit platelet aggregation, whereas (b) lower concentrations enhance it. PGE2 activates four types (PGE2 type 1 or EP1, PGE2 type 2 or EP2, PGE2 type 3 or EP3, and PGE2 type 4 or EP4) of its G protein-coupled receptors. Each of these receptors has a distinct pharmacological signature and intracellular signal transduction. Stimulation of EP3 receptors results in elevation of free intracellular calcium levels, whereas stimulation of EP2 and EP4 receptors usually decreases intracellular calcium levels due to increase in intracellular cAMP levels through the activation of Gas (growth arrest-specific) protein [41]. It may further be mentioned that except EP2 receptor, the mRNA for EP1, EP3, and EP4 receptors is present in human platelets [41].

The activation of the EP3 receptor by leading to inhibition of the increase in cAMP increases the mobilization of calcium and elevates P-selectin expression in platelets ascribing the proaggregatory effect of PGE2. Lacking of this receptor has been found to show an increased bleeding tendency and a decreased susceptibility to thromboembolism [42], but PGE2 produced by atherosclerotic plaques can further facilitate arterial thrombosis via EP3 [43]. Interestingly, no defects in the EP3 receptor gene have been found in humans [7]. The platelet aggregation, calcium mobilization, and upregulation of P-selectin are found to be inhibited by the selective EP4 agonist, ONO AE1-329. Additionally, the EP4 antagonists, GW627368x and ONO AE3-208, have been found to repeal the inhibitory effect of ONO AE1-329 on platelet aggregation [44]. Thus, the EP4 receptors might play an important role mediating the inhibitory effect of PGE2 in the control of hemostasis by balancing out the proaggregatory effect of EP3 receptors. EP4 agonists might constitute a novel class of antithrombotic agents and also might be clinically useful in those cases where aspirin or ADP antagonists are not warranted or are insufficient, as the EP4 activation enhances the inhibitory effect of aspirin [44].

5.2 Prostaglandin I2 (PGI2) or prostacyclin receptor

Prostaglandin I2 (PGI2) or prostacyclin is a derivative of arachidonic acid, released by vascular endothelial cells (**Table 1**). It is an effective (a) vasodilator, (b) platelet aggregation inhibitor, and (c) moderator of vascular smooth muscle cell proliferation-migration-differentiation (anti-atherosclerotic). It acts through a specific membrane-bound receptor, the prostacyclin receptor (IP receptor). The IP receptor belongs to the prostanoid family of GPCR. The receptor (37–41 kDa, depending upon different states of glycosylation) are class A rhodopsin-like GPCR. The glycosylation of the extracellular domain is necessary for (a) ligand binding, (b) receptor activation, and (c) membrane localization. A number of serine residues (S328 and S374) are thought to be phosphorylated by GPCR kinases or second-messenger-activated kinases (PKC and PKA) in the cytoplasmic domain. It might play a potential role in either agonist-induced phosphorylation or kinase-mediated receptor desensitization [45]. Cyclopentane ring and side chains are the two structural features of prostaglandins. Among these, the side chains are found to be recognized by their receptor to stabilize ligand binding, and the binding pocket of the receptor can accommodate the cyclopentane rings (PGI2, PGE1, and PGE2). The IP is the most common to be associated with the Gas subunit of the heterotrimeric G-protein. Upon receptor activation, it has been found to catalyze the formation of the second messenger, cAMP by stimulating the membrane-bound adenylyl cyclase [7].

The discussion with GP, TX, and PG lead us to discuss with thrombin receptor as stimulation of thrombin activates the thrombin receptors *in vivo*.

6. Thrombin receptors

As explained earlier, platelet activation by thrombin partially depends on GPIb-IX-V, but is primarily assured by two protease-activated receptors (PAR), i.e., PAR-1 and PAR-4. Binding of thrombin (immobilized, proteolytically inactive) to GPIb induces platelet adhesion as well as spreading and secretion [19], being an enhancer (GPIb) of the thrombin response (see Section 2.1). PAR-1 and PAR-4 (**Table 1**) are activated by a unique irreversible proteolytic cleavage (within the first extracellular loop exposing an N-terminus) by serving as a tethered ligand to GPIb. Short synthetic peptide mimetic of the N-terminus sequences, corresponding to the new N-terminus, reproducing most of the action of thrombin on platelets upon cleavage by thrombin (SFLLR for PAR-1 and GYPGQV for PAR-4) can activate these receptors directly. This dual-receptor (PAR-1 and PAR-4) signaling for thrombin implies that PAR-1 is the primary mediator that activates platelets at low concentration, whereas PAR-4 as a back-up receptor is found to be activated at higher thrombin concentrations. For sustained optimal platelet responses to thrombin, the qualitative differences in the dynamics of PAR-1 and PAR-4 activation might be relevant. The signaling for PAR-4 at high thrombin concentrations acts very slowly, as the PAR-4-mediated Ca^{2+} mobilization is found to be slower and more prolonged than that of PAR-1, and also, this activity terminates more slowly. Interestingly, there is no such report with congenital deficiencies of PAR receptors in any individuals. The pharmacological invention of PAR-1 inhibitors will be very helpful to prevent the platelet-dependent thrombosis in the first stage [46].

On the surface of platelet membrane a type of kinase receptors are present which are specific to its ligand and associated with integrin receptors to act with, known as ephrin kinase receptor.

7. Ephrin (Eph) kinase receptors

Eph kinases are receptors expressed on the surface of cells. It activates in response to binding with Eph receptor-interacting proteins, ephrins. Eph kinases are known to be a member of receptor tyrosine kinases subfamily with an extracellular ligand binding domain and an intracellular tyrosine kinase domain (**Table 1**). The EphB is distinguished from the EphA subfamily by an insertion within the extracellular domain that helps to define the ligand preferences for the receptor. EphrinA to EphA interaction (**Table 1**) occurs typically with higher affinity than ephrinB (to EphB) interactions. This may be due to the fact that EphAs bind via a "lock-and-key" mechanism with little conformational change with less energy, in contrast to EphBs which utilizes an "induced fit" mechanism with a greater amount of energy to alter the conformation of EphBs for binding to ephrinBs [47]. The Eph kinase and ephrin interactions on adjacent cells play a pivotal role in neuronal patterning and vasculogenesis. Eph subtypes of EphA4, EphB1, and also ephrinB1 are found to be expressed by the human platelets. In both resting and activated platelets, the EphA4 is constitutively associated with $\alpha IIb\beta 3$. Fine tuning between the Eph and ephrin is very much essential as clustering of either EphA4 or ephrinB1 causes platelet adhesion to immobilize fibrinogen, whereas by blocking this (Eph/ephrin) interaction, the clot retraction can be hampered. This may cause platelet aggregation inhibition at low agonist concentrations and may form smaller thrombi on collagen-coated surfaces during normal conditions of arterial flow. This can develop premature disaggregation. It acts partially due to the ability of ephrin B1 to activate Rap1 (a Ras family member)-mediated signaling which supports platelets activation, especially the integrin in platelets [48].

There are also purinergic (P) receptors, present on the platelet membrane, which are basically ADP or ATP dependents.

8. Purinergic receptors

These purinergic receptors reside in coupled form, that is why it represents as P2. These P2 (purinergic coupled) receptors are of mainly three types—P2Y1, P2Y12, and P2X1 (**Table 1**).

8.1 P2Y1

The P2Y1 receptors (42-kDa, contain 373 amino acid residues) are widely distributed in many tissues (heart, blood vessels, smooth muscle cells, neural tissue, testis, prostate, and ovary) including platelets. About 150 P2Y1 receptor-binding sites are expressed per platelet, and it is also abundantly represented in membranes of α-granules and elements of the open canalicular system. The P2Y1 receptor is absolutely required for ADP-induced platelet aggregation. ADP is a more potent agonist than ATP, and its 2-methylthio derivatives are more potent than the parent compounds. ATP is a partial agonist for the P2Y1 receptor, and at the low levels of receptor expression, it acts as an antagonist. Overall, P2Y1 accounts for about 20–30% of the total ADP-binding sites on the platelet surface [49]. At the P2Y1 locus, a common genetic variant (dimorphism, 1622AG) exist which is associated with platelet reactivity to ADP. However, this can partly explain the interindividual variation in platelet's response to ADP and may have clinical implications in relation to the thrombus formation [7].

8.2 P2Y12

The purinergic P2Y12, the Gi-coupled platelet receptor, exists in platelets, smooth muscle cells, endothelial cells, and glial cells. This surface protein is expressed from the chromosome 3q21–25 containing 342 amino acid residues which includes extracellular Cys residues at four different locations (17, 97, 175, and 270, respectively). The Cys 97 and Cys 175 are found to be linked by a disulfide bridge, and this link is important for receptor expression in platelets. P2Y12 receptor exists predominantly on the platelet surface among the purinergic receptors as homo-oligomers placed in lipid rafts. The active metabolite of clopidogrel (which covalently inhibits P2Y12) application as treatment disrupts the homo-oligomers into nonfunctional dimers and monomers which are sequestered outside the lipid rafts [7]. ADP and its analogs (e.g., 2-methylthio-ADP and N-methanocarba-2-methylthio-ADP) stimulate the P2Y12 receptor, while ATP and its triphosphate analogs act as antagonists to it [7]. P2Y12 plays a central role for ADP in platelet function. The congenital P2Y12 defective patients display a mild to moderate bleeding diathesis with the conditions of mucocutaneous bleedings and postsurgical and posttraumatic excessive blood loss. Any defects of P2Y12 can also be thought of even when high concentrations (10 mM) of ADP is unable to induce full, irreversible aggregation of platelets [49].

8.3 P2X1

P2X1 is a widely distributed ligand-gated ion channel, highly expressed in human megakaryocytes and platelets. It is well known that ATP is the physiological agonist and ADP is an antagonist. Platelet dense granules release ATP upon its activation with an ion channel (cationic and/or anionic) present on platelets, i.e., P2X1. During platelet preparation under *in vitro* condition, a rapid desensitization of the P2X1 receptor occurs which made this receptor unnoticed. The P2X1 gene lies on the chromosome 17p13.2. It encodes 399 amino acids which are organized into two transmembrane domains (TM1 and TM2). These are separated by a large extracellular domain containing 10 cysteine residues. Three molecules of ATP bind to the extracellular domain of P2X1 and trigger conformational changes. This results in the opening of a cationic pore for monovalent and divalent cations (e.g., Ca^{2+}, Na^{+}, and K^{+}) allowing rapid changes in the membrane permeability. P2X1 receptor activation triggers transient shape change of a platelet from a discoid to spherical shape; furthermore, in presence of low concentrations of other platelet agonists, it (activation of P2X1) amplifies platelet responses. The P2X1 receptor contributes equally to low and high levels of thromboxane A2 receptor activation. More significantly, P2X1 receptor activation has been shown to be essential for enhanced platelet adhesion and thrombus formation under high shear rates [50]. P2X1 is distinguished as a potential new drug target for antithrombotic therapy, especially for the mild long-term risk management. As inhibition of P2X1 causes mild effects on different platelet function, it (P2X1) seems to be a so-called "safe" target clinically [51].

As mentioned earlier (see Section 5.1), Gas receptors also have some important role in the stabilization of platelet aggregation to maintain internal hemostasis.

9. TAM receptor and Gas6

Gas6 has its interactions with Tyro 3, Axl, and Mer (TAM) tyrosine kinase receptors (**Figure 1**). It is a vitamin K-dependent protein implicated in cell growth, adhesion, and migration, through its TAM receptor. These three related protein receptors (TAM) were cloned (in 1991) as orphan receptors and have been found to be widely expressed in the vertebrate nervous system [7]. They (TAM receptors) have two N-terminal immunoglobulin domains, followed by two fibronectin-III-like domains to mediate binding to the ligand. These domains are found to be attached to an intracellular tyrosine kinase domain, via a single-pass α-helical transmembrane domain. The functional receptors form both the hetero- and homo-dimers, which is common with other receptor tyrosine kinases (**Table 1** and **Figure 1**).

It (Gas6) is found in plasma and platelet granules, but in human, it is present predominantly in plasma. It is secreted upon platelet activation, as mentioned earlier. Deficiency in Gas6 or one of its receptors (TAM) in mice has shown abnormal platelet responses to agonists and prevents thrombosis, suggesting a major role of this Gas6-TAM receptor-coupled axis in thrombus formation as well as in vascular wall homeostasis. The role of Gas6 in human platelet function has been clarified with the fact that the Gas-6 reinforced αIIbβ3 integrin with the outside-in signaling by the activation of PI3K and Akt and therefore clot retraction by promoting the β3 phosphorylation [52]. These effects constitute in fact an enhancement and perpetuation of the thrombus-stabilizing role of ADP. The inhibition of Gas6 signaling has been proposed as an attractive target for novel antithrombotic drugs [53].

Many receptors are known of transmembrane family. There is also a unique tetraspan transmembrane receptors on the platelet membrane, known as tetraspanin receptors.

10. Tetraspanin receptors

Tetraspans, mean four transmembrane domains, called as tetraspanins belong to the transmembrane 4 superfamily (TM4SF) receptors. Usually, tetraspanins are found to act as scaffolding proteins. It has intracellular (N- and C-) termini and two extracellular domains (EC1 and EC2), arranged in a way of one short and one long with typically 100 amino acid residue long loop. Its EC2 domain with four or more cysteine residues are its main feature, among which two are in a highly conserved "CCG" pattern. It can anchor multiple proteins to one area of platelet cell membrane [7].

10.1 CD151

The CD151, a tetraspanin superfamily member, previously termed as PETA-3/SFA-1 has been found to express broadly in hematopoietic, vascular, and immune compartments, and especially abundant in cardiac muscle, endothelia, epithelia, megakaryocytes, smooth muscle, and the platelets. It is found to functionally link with the integrin trafficking, cell migration, cancer metastasis, neurite outgrowth, hemidesmosome formation, vascular morphogenesis, wound healing, immune responsiveness, and hemostasis. CD151 has been found to appear to regulate fibrinogen-binding proteins (e.g., integrin αIIbβ3) (**Table 1**). In addition, the absence of CD151 *in vivo* leads to smaller, unstable thrombi formation [54].

10.2 TSSC6

The TSSC6 or tumor-suppressing subchromosomal transferable fragment cDNA 6 also called as pan-hematopoietic expression (Phemx). It is a member of tetraspanin superfamily. Its C-terminal cytoplasmic domain is relatively large (33 amino acids in mouse and 99 amino acids in human) than the other members of the tetraspanin superfamily [55]. TSSC6 may modulate hematopoietic cell function specifically when expressed in hematopoietic organs and tissues. It is expressed on the surface of murine platelets and is upregulated by thrombin stimulation. The secondary stability of arterial thrombi formation (by regulating integrin αIIbβ3 outside-in signaling events) has been found to affect upon vascular injury during the lack of platelet TSSC6 receptors *in vivo* (**Figure 1**) [55]. The proliferation of T lymphocytes has also been observed in the TSSC6-deficient mice, due to the increase in interleukin 2 production following T-cell receptor stimulation, providing a clue to the negative regulation of peripheral T-lymphocyte proliferation by TSSC6.

10.3 CD36

CD36 (80–90 kDa) is known to be a scavenger receptor. This CD36 along with other receptors and their corresponding subunits (GPIIIb, GPIV, GP88, FAT, SCARB3, or PASIV) are expressed on the surface of the platelets and other cells (e.g., monocytes, endothelial cells, smooth muscle cells, and cardiomyocytes). Approximately 10,000–25,000 molecules are present in a single platelet. Its gene is located on the chromosome 7, in case of human.

It consists of a single peptide chain (consisting of 474 amino acids). Its two transmembrane domains (one near the N-terminus and the other near the C-terminus) are configured like a "hairpin-like" structure. The domains of CD36 are separated by a large, glycosylated extracellular loop. It (CD36) was initially described as a collagen receptor of types I and III on platelets, but later, it has been found that it is not a primary collagen receptor as its binding to the nonfibrillar type V collagen has been documented [56]. The interaction of CD36 with its various ligands (e.g., thrombospondin 1 or TSP1, long chain fatty acids, oxidized phospholipids or oxPL, and oxidized low-density lipoprotein or oxLDL) is found to modulate the platelet activation (**Table 1**). TSP-1 is found to promote platelet aggregation through the modulation of an inhibitory signaling pathway. TSP-1 binding to its receptor CD36 prevents cAMP/protein kinase A (PKA) signaling. Indeed, TSP1 triggers CD36-dependent signals that reduce platelet sensitivity to PGE1. It (CD36-dependent signaling) diminished its (PGE1) ability to inhibit platelet aggregation and arrest under conditions of flow [57]. Other CD36 ligand, oxLDL, formed during hyperlipidemia and atherosclerosis can also activate platelets in a CD36-dependent manner, as stated in TSP1. The level of platelet CD36 surface expression is highly variable among individuals within general population with inheritance of specific genotypic polymorphisms at the CD36 locus (**Figure 1**) [58].

10.4 TLT-1

Another member of this tetraspan superfamily is the triggering receptors expressed in myeloid cells or TREMs, which are found to be involved in the activation of various cell types of the innate immune system. It includes platelets, monocytes, macrophages, microglia, and neutrophils. The family is characterized by a single V-set immunoglobulin (Ig) domain, a short cytoplasmic tail, and a charged residue in the transmembrane domain. TREM-like transcript-1 (TLT-1 or TREML-1) is a type I single Ig domain orphan receptor. It is specific to platelet and megakaryocyte alpha-granules. It only relocates to the platelet surface upon platelet stimulation. Its longer cytoplasmic tail carries a canonical ITIM (immunoreceptor tyrosine-based inhibition motif) which is capable of becoming phosphorylated and of binding to the Src homology-containing protein tyrosine phosphatase-1 (SHP-1). It (ITIM) can identify TLT-1 as the only putative inhibitory member of the TREM cluster. During storage, the ability of anti-TLT-1 scFv (single chain variable fragment) to block aggregation of washed platelets suggested that TLT-1 facilitates thrombosis by interacting with ligand(s) in activated platelets. TLT-1 acts in collaboration with αIIbβ3 to facilitate fibrinogen/platelet interactions and/or higher order platelet aggregation following the same signaling mechanism (**Figure 1** and **Table 1**) [59, 60].

10.5 PEAR1

The PEAR1 or platelet endothelial aggregation receptor-1 (150 kDa) is known as multiple epithelium growth factor 12 (MEGF12) or Jedi-1. Interestingly, "Jedi" is not a scientific name here, which is being used by the scientists. It is a myth to describe the power and devotion of a knight, known as Jedi knights who respect all life by defending and protecting those who cannot do encounter for themselves, and in any altercations, they remain ready to encounter and fight only in self-defense and for the defense of those they protect. Based on this myth's symbolic importance and strength of PEAR-1 receptor in the maintenance of homeostasis, it was named so.

It is a transmembrane protein of the MEGF-like domain protein family. It is mainly expressed in platelets, endothelial cells, and also in satellite glial cell precursors. During development, it is necessary for the clearance of apoptotic neurons via phagocytosis in the embryonic dorsal root ganglia (DRG). PEAR1 is composed of (a) an extracellular Emilin domain (EMI domain), (b) 15 extracellular EGF-like repeats, and (c) multiple cytoplasmic tyrosines and pralines, (d) intracellular domain structure, containing 5 proline-rich domains and an NPXY motive (serving as a phosphotyrosine-binding site and an internalization signal). During platelet aggregation, PEAR1 is phosphorylated at Tyr-925 and Ser-953/1029 in an αIIbβ3-dependent manner of signaling mechanism, as described previously. The PEAR1 has been hypothesized as a platelet-platelet contact receptor, due to its (PEAR1) αIIbβ3-independent phosphorylation [61]. The high-affinity immunoglobulin E receptor subunit α (FcεR1α) has been found as PEAR1 ligand (**Table 1**). PEAR1 promoter-region variant (rs2768759) was associated with increased aggregation in PRP, most strongly in response to epinephrine, in both pre- and post-aspirin treatment conditions [62]. Increased expression of PEAR1 might be an important cause of hyperactivity [62] and genetic variation within PEAR1, particularly rs41299597, seems to lead to an increased membrane expression of PEAR1 in activated platelets and elevated responsiveness to GPVI ligands [63]. A genome-wide meta-analysis linked the minor allele of the PEAR1 SNP (rs12566888) to a drop in aggregation response toward ADP and epinephrine in the European and African-ancestry sample [64].

Cell-cell adhesion is very important in the process of thrombus formation and obviously to extend the formation into proper stability until the proper clot forms. In this process, a specific adhesion molecule and its specific ligand have been found to play an important and interesting role to maintain hemostasis. P-selectin is a member of the selectin family of adhesion molecules. P-selectin glycoprotein ligand-1 (PSGL-1) on the plasma membranes of neutrophils or monocytes, which are key effector cells of the innate immune system, binds to P-selectin translocated to the surfaces (**Figure 1** and **Table 1**) of the inflamed endothelial cells or activated platelets [65].

11. P-selectin

Selectins are a family of cell adhesion molecules (CAMs). It is also known as clusters of differentiation 62 or CD62. It is present in endothelium (as E-selectin, 58.6 kDa), leukocyte (as L-selectin, 30 kDa), and platelets (as P-selectin). P-selectin has a molecular weight of 86 kDa based on the prediction from its cDNA, but from reducing SDS-PAGE, it is about 140 kDa. The primary ligand for P-selectin is P-selectin glycoprotein ligand-1, known as PSGL-1 which is found constitutively on all leukocytes (**Table 1**). The transient interactions between P-selectin of activated platelets and PSGL-1 of leukocytes allow them to roll along the venular endothelium. During this activation of the coagulation cascade, the formation of a fibrin network is found to be a critical event in thrombus stability (**Figure 1**). A laser injury-induced thrombosis is found to express a low level of tissue factor (TF) in mice. It has also been shown that this fibrin formation depends on the monocyte-derived TF, carried by microvesicules, with minimal contribution of vessel wall TF. These microvesicules are captured onto the thrombus through the interaction between P-selectin and PSGL-1 (as mentioned before) present on microvesicules, hence delivering TF to the growing thrombus. Based on this observation, studies with mice, deficient in either PSGL-1 or P-selectin display thrombi with little TF and reduced thrombin generation, resulting in hampered thrombus size [7, 65]. Elevated plasma P-selectin (normal value 100 ng/ml in man) is a major predictive factor of cardiovascular events related to platelet turnover and its activation as well as function. So, the increase in P-selectin expression is expected to develop the artery diseases of peripheral tissues, stroke, and even the acute myocardial infarctions [7].

Being a dynamic process, a state of surface is preferable in the arterial thrombus formation to limit thrombus growth passively. In this context, it is not unreasonable to mention here that the essential roles of nitric oxide (NO) and PGI2 in the negative regulation of platelets to prevent uncontrolled thrombosis have been well established [7]. However, the inhibitory role of various receptors with and without immunoreceptor tyrosine-based inhibition motif (ITIM) domain for active thrombus formation is well recognized [7, 65]. It is very important to limit the thrombus formation within the blood vessels. These receptors are now in focus of our discussion.

12. ITIM-containing receptors

Immunoreceptor tyrosine-based inhibition motifs or ITIMs are defined by a consensus sequence of (L/I/V/S)-X-Y-X-X-(L/V); ITIM-containing receptors are found in pairs, separated from each other by 15–30 amino acid residues, were originally identified by their ability to inhibit signaling by its activation counterpart (immunoreceptor tyrosine-based activation motif or ITAM) receptors. The ITAM and ITAM-like receptors, GPVI and CLEC-2 in the presence of thrombin (a GPCR agonist), cause mild inhibition of platelet activation by the platelet endothelial cell adhesion molecule or PECAM-1. The action of this inhibition by the thrombin, GPVI, CLEC-2, and PECAM-1 is similar to that of G6b-B in which it (G6b-B) can inhibit the SFK (Src and Syk) signaling to prevent unwanted platelet activation. This signaling mechanism also shows the inhibition of platelet activation by the GPVI-specific agonist (collagen-related peptide) and ADP (an agonist of GPCR) (**Table 1** and **Figure 1**).

12.1 PECAM-1

PECAM-1 (or CD31, molecular weight 130 kDa), the adhesion molecule, bears a pivotal role in the negative regulation of platelet aggregation by inhibiting the platelet activation. PECAM-1 is expressed on the cell surface of hematopoietic and immune cells, which include platelets, neutrophils, monocytes, megakaryocytes, natural killer cells, some T cells, and on endothelial cells, particularly at the borders of the adjacent cells. PECAM-1 is a member of the Ig superfamily (like GPVI receptor), consists of (a) six extracellular Ig domains, (b) transmembrane domain, and (c) cytoplasmic tail. The cytoplasmic domain contains an ITIM, which becomes phosphorylated upon stimulation by homophilic interactions and/or clustering. PECAM-1 is found to ease the recruitment of tyrosine, serine/threonine, sometimes possibly the lipid phosphatases, and consequent kinase-dependent signaling inhibition to inhibit the platelet activation by attenuating the thrombus formation and thrombin-mediated platelet activation (by means of negative regulation) involving GPVI and GPIb (**Table 1**) [66]. PECAM-1 is an efficient signaling molecule in platelet and is capable of exhibiting both outside-in and inside-out signaling. PECAM-1 is also implicated in numerous other biological functions including apoptosis, platelet aggregation, thrombosis, and angiogenesis.

12.2 G6b-B

Among the novel plasma membrane proteins (identified via proteomics study), the immunoglobulin superfamily member G6b is one of them. It consists of 241 amino acids (26 kDa). It is found to undergo extensive alternate splicing. In stimulated platelets, G6b-B undergoes tyrosine phosphorylation in association with the Src homology-2 (SH2) domain-containing phosphatase (SHP-1). It suggests its importance to play a novel role in limiting platelet activation. Only the G6b-B is found to have both (a) a transmembrane region and (b) two ITIM. This ITIM supports binding to the two SHPs, i.e., SHP1 and SHP2. Heparan sulfate has recently been found as a ligand for G6b-B receptors (**Table 1**). The second ITIM, with a slightly different sequence (TXYXXV), is located around 20 amino acids downstream of the first ITIM of G6b-B [67]. In addition, G6b-B is encoded by a gene that is variously called G6b, C6orf25, or MPIG6B. It is a platelet and megakaryocyte-specific receptor.

12.3 VPAC1

VPAC1 is the vasoactive intestinal peptide (VIP)/pituitary adenylate cyclase-activating peptide (PACAP) receptor 1. The PACAP is a neuropeptide of the VIP, a member of secretin/glucagon superfamily. The PACAP receptor (vasoactive intestinal peptide/pituitary adenylate cyclase-activating peptide receptor 1) or VPAC1 in platelets is coupled to adenylyl cyclase activation. The VPAC1 and VPAC2 are found to couple with the G-protein (Gs) resulting in the stimulation of cell adenylyl cyclase. The VPAC1, together with the VPAC2 receptor subtype, mediates a large array of VIP or pituitary adenylate cyclase activating peptide actions on different physiological functions including exocrine secretions, release of hormones, relaxation of muscles, metabolism, growth control of fetuses, and embryonic brain development. Patients with severe mental retardation have found to have a bleeding tendency with mild thrombocytopenia. Increased basal cAMP level in platelets is providing a basis for the reduced platelet aggregation. Megakaryocyte-specific transgenic overexpression of PACAP consequently increased the PACAP release from platelets to reduce the platelet activation and thereby prolongation of the bleeding time. In the management of arterial thrombosis and arterial bleeding, the therapeutic potential of PACAP is now considering clinical practices as inhibitor of thrombus formation [68].

13. Toll-like receptors (TLRs)

These are transmembrane proteins consisting of a lipoprotein receptor-related protein (LRP) extracellular domain, a transmembrane region, and a Toll-IL-1R domain, present on the surface of platelets. These exist in smaller amounts than GPIb-IX-V complex. Human platelet TLRs are similar to Toll receptor in *Drosophila* with four types (TLR-1, TLR-2, TLR-4, and TLR-6). However, platelets and megakaryocytes express mRNA and/or protein for different TLRs (e.g., TLR-1, TLR-2, TLR-3, TLR-4, TLR-6, TLR-7, TLR-8, and TLR-9) that detect and bind viral components and nucleic acids at the cell surface [69]. These TLRs (**Table 1**) play an important role in innate immunity by their ability to identify the products of bacteria, viruses, protozoa, and fungi, and important for their clearance (**Table 2**). After identification of these products, TLRs activate intracellular signaling pathways to

Receptors	Viruses
αvβ3	Hantaviruses, coxsackieviruses A9 and A16, human adenovirus type 2, echovirus 9, and human parechovirus
αIIbβ3	Human parechovirus
α 2β1	Echovirus 1 and rotavirus
DC-SIGN	Lentivirus, HIV, and Ebola virus
Axl	Lassa fever virus (LASV)
Tyro3; CCR-3 and -4; CXCR-1, -2, and -4; CLEC-2; DC-SIGN	HIV
GP-VI	HCV
CR2	EBV
α2β1	Rotavirus
α2bβ3 (GPIIbβ3a)	Adenovirus

CLEC-2: C-type lectin-like type II transmembrane receptor; CR: complement receptor; CCR: C-C chemokine receptor; CXCR: C-X-C chemokine receptor; DC-SIGN: dendritic cell-specific intercellular adhesion molecule-3-grabbing; non-integrin; GP-VI: glycoprotein VI. Other details of platelet receptors are same as presented in ***Table 1****. (This table is adopted from Seyoum et al. [69].)*

Table 2.
Different classes of platelet receptors for viruses.

induce an inflammatory response [70]. TLR-4 promotes platelet-neutrophil interaction and also causes activation of the neutrophils [71]. However, the involvement of FcγRIIA and serotonin may not be ignored [72].

14. Serotonin receptors

Serotonin (5-hydroxytryptamin) receptor of 2A type or 5-HT2A and 5-HT3A has also been found on platelet membrane [73, 74]. Platelet activation has been found to release stored serotonin from dense granules by amplifying release reaction and thereby promoting platelet aggregation. However, serotonin itself does not cause platelet aggregation but enhances platelet aggregation, induced by its other agonists (e.g., ADP and thrombin) [75]. It has been depicted that the interaction of 5-HT2A with serotonin initiates calcium signaling. The 5-HT3 receptor has been found to increase immunoreactivity with the platelet activation (with ADP and thrombin receptor activating peptide, TRAP). Serotonin is also known to cause vasoconstriction of the blood vessels with damaged endothelium and promotes thrombus formation. It has the ability to attach to a large number of substrates including fibrinogen, vWF, thrombospondin, and fibronectin [76]. The release of serotonin precedes the neutrophil contribution to shock, consistent with the reported role of serotonin in neutrophil activation (**Table 1** and **Figure 1**). The vasodilation was also present in the absence of neutrophils, suggesting that the platelets orchestrate neutrophil activation and endothelial cell functions through serotonin and this event can occur independently critically with the FcγRIIA and serotonin [72].

15. Leucine-rich repeat receptors (LRRs)

The LRR is a protein structural motif with a repetition of 20–30 amino acids which are rich in leucine, a hydrophobic amino acid. It is involved in the formation of protein-protein interactions. This LRR includes GPIb-IX-V complex (see Section 2) and TLRs (see Section 13) of platelets. Though these receptors have their own class, they also have similarities or common factors between them (**Table 1**).

16. Complement receptors

Several types of complement receptors (CRs) are also expressed by the platelets (e.g., CR2, CR3, CR4, C3aR, C5aR, gC1qR, and cC1qR) [77]. These complements act as receptor for pathogens (**Table 2**) and implement multiple functions with both direct and indirect antimicrobial host defense (including cell lysis, opsonization, and chemotaxis) [69].

17. DC-SIGN

Recently, it has been found that the platelet granules can express the dendritic cell-specific ICAM3-grabbing non-integrin (DC-SIGN) receptor (**Figure 1** and **Table 2**), like other cells which are used for the interaction with HIV-1 [78]. Platelet surface DC-SIGN receptors or enhancement of the receptor FcγII play an important role in binding of platelets with Dengue virus (DENV). DC-SIGN and heparin sulfate proteoglycan are receptors for DENV [69]. DENV infection leads to thrombocytopenia by increasing phagocytosis of DENV-induced platelets apoptosis by macrophages via a phosphatidylserine-recognizing pathway [69].

This detailed description of different platelet receptors will help us to understand the molecular changes of platelet receptors in different conditions.

18. Molecular changes of platelet receptors in different conditions

18.1 During platelet storage

In the cases of thrombocytopenia and bleeding complications, transfusion of platelet concentrates (PCs) is one of the most important therapeutic approaches and its management formula. Regardless of preparation techniques, stored platelets gradually experience inevitable deleterious changes called as platelet storage lesion (PSL). It may lead to a progressive structural and functional damage of platelet adhesion receptors from the time of platelets isolation till the transfusion to a recipient. It is induced mainly by either reversible or irreversible increases in the basal levels of platelet activation, and with the irreversible changes in platelet morphology and function the most important phase of PSL is associated with. These changes of platelet are significantly initiated with platelet α granules release (as mentioned in Sections 11 and 12) and its conversion from pro-aggregatory status to pro-inflammatory phenotypes which is identified by the P-selectin and CD40L expression [86]. The platelet activation increases the intracellular Ca^{2+} levels and acts as the main modulator of signaling events. This results in platelet receptor ectodomain shedding and membrane loss due to microparticulation. Both of these phenomena (ectodomain shedding and membrane loss) can cause progressive loss of platelet adhesive receptors during platelet storage [4, 87] and may affect its proper function required for therapeutic uses. During the storage of platelet concentrates (PCs), the GPIbα expression levels have been found to be decreased [77, 78]. The continuous shedding can lead to a significant decrease in GPIbα expression in 5-day-stored platelets [86]. More old platelets with the lower GPIbα expression may show less functionality after transfusion. The metalloproteinase-dependent loss of surface GPIbα plays a role in the clearance of aged platelets from the circulation. In older PCs, these receptors with higher levels of shedding could also be associated with a rapid clearance of transfused platelets leading to reduced platelet recovery and survival. In addition to this, the GPVI receptor expression also showed to be modulated during storage. A negative correlation between GPVI expression and shedding was also observed, a finding that verifies the main role of ectodomain shedding in the modulation of GPVI expression. There is a direct correlation between the shedding levels of P-selectin and GPVI during platelet storage. It suggests that this adhesion receptor is a valid marker of PSL [87]. The levels of soluble GPVI have been correlated with soluble P-selectin in patients with acute coronary syndrome and/or with acute ischemic stroke [88]. Considerable shedding of GPVI and its association with decreasing expression of this receptor on the surface of stored platelets can affect GPVI-dependent platelet function during storage. There is a direct correlation between this observed decreased adhesive capacity with the increasing levels of GPVI shedding during storage [4].

This enucleated blood component (platelet) may have a chance to interact with or interacts definitely under different pathophysiological conditions. Now, we will focus on those pathophysiological changes in relation to the platelet receptors.

18.2 During different pathophysiology

18.2.1 Inflammation

Platelets do not only play a role in thrombus formation but are also important in atherogenesis, and deficiency of an individual receptor or ligand has been found to be compensated by other receptors [89–92]. The inhibition of platelet adhesion in atherosclerosis-prone mice impedes the development of atherogenesis in it (these models of study) [93]. Platelet adhesion receptors discussed above in the context of thrombosis also support interactions between platelets, endothelial cells, and leukocytes involved in other vascular processes [94, 95]. Activated platelets with a potential mechanism adhered (by GPIbα binding) to the vessel wall with the support of leukocyte adhesion (with integrin, αMβ2). The GPIbα-binding site of the GPIb-IX-V complex similarly involves in the insertion of αM I domain of the αMβ2 to the vWF, homologous to the vWF-A1 domain [8]. Another mechanism involved here is the interaction between the platelet P-selectin and leukocyte PSGL-1 binding, as mentioned previously [9], and finally, a link between platelets and endothelial cells forms by the fibrinogen or vWF via the engagement of integrins on both platelets (such as αIIbβ3) and endothelium (such as αvβ3) [96]. Understanding of the involvement of the networking of these platelet-specific receptors in thrombosis pathophysiology *in vivo* is yet to confirm.

18.2.2 Bernard-Soulier syndrome

The Bernard-Soulier syndrome (BSS) is the defect in the three GPIb encoding genes (as mentioned earlier), which gives rise to a serious bleeding diathesis, accompanied by anomalies in platelet morphologies, especially giant platelets. It is a rare hereditary thrombocytopathy, first described in 1948 by Jean Bernard and Jean-Pierre Soulier, two French hematologists, in a young male patient who had severe mucocutaneous bleeding, prolonged bleeding time with normal platelet count, and abnormally large platelets (macrothrombocytopenia) [90]. In view of these defects, the disorder was named as “Dystrophie thrombocytaire-hémorragipare congénitale” (hemorrhagiparous thrombocytic dystrophy) [97]. In most cases, bleeding symptoms manifest rapidly after birth or during early childhood. Clinical manifestations usually include purpura, epistaxis, gingival bleeding and menorrhagia (menstrual periods with abnormally heavy or prolonged bleeding), and more rarely gastrointestinal bleeding and hematuria (presence of blood in a person’s urine). Severe bleeding episodes are associated with trauma and surgical procedures (e.g., tonsillectomy, appendectomy, splenectomy, during dental extractions), gastric ulcers, and menses. However, indeed the severity and frequency of bleeding vary between individuals [21]. Ultrastructural studies of affected platelets show a dilated open canalicular system, prominent dense tubular system, and vacuolization. It is rare with a reported prevalence of 1 in 1,000,000. Patients have a prolonged bleeding time, thrombocytopenia, and larger platelets than the normal individual due to defective thrombopoiesis in GPIb defective megakaryocytes [98]. Few mutations were reported that cause a gain-of-function in the GPIbα chain, leading to the so-called platelet-type GPIB [99], showing a phenotype similar to that of certain subtypes of von Willebrand disease [100]. The syndrome, as an autosomal recessive trait, is found to be transmitted with an underlying defect of deficiency or dysfunction of the GPIb-V-IX complex, required for normal primary hemostasis. The GPIb-V-IX complex binds to vWF, allowing platelet adhesion and platelet plug formation at sites of vascular injury (as mentioned in Section 2.1) (**Table 1**) [21].

18.3 Aging and aging-related pathophysiogical conditions

During aging, the platelet receptor expression levels have also been investigated and found that the platelet responsiveness to collagen decreases with age with an aging-dependent decrease in GPVI-dependent platelet activation [101]. Adhesion receptor levels on nucleated thrombocytes have also been found to decrease during aging in correlation with the decrease in participation in thrombus formation. The younger thrombocytes contain more adhesive receptors with a higher propensity to form the thrombi than the aging counterpart [102]. The mechanism of platelet shedding in regulation of aging-related changes in adhesion receptor levels has not been demonstrated yet [103]. Aging with the different biochemical alterations due to alteration in cellular microenvironments positively associates with the downregulation of different receptors, its consequent signaling pathways, biochemical cascade mechanisms, formation of abnormal proteins, its related neurodegeneration, immunosuppression, susceptibility to viral infection (e.g., COVID-19), etc. and negatively associates with the survival from the diseased condition.

18.3.1 Thromboxane alterations

Thromboxane (A_2 and B_2) is produced from arachidonic acid through endoperoxidase by the cyclooxygenase and thromboxane synthase enzyme activity, respectively [104]. It plays a pivotal role in platelet aggregation *in vivo*. It has been found that thromboxane production is enhanced in diabetic subjects resulting in platelet aggregation, which further provides a higher risk of cardiovascular disease [104]. A number of risk factors have been proposed to elevate CVD risk in type 1 diabetes mellitus (T1DM) patients, including hyperglycemia, dyslipidemia, inflammation, oxidative stress, and genes among others. A significant contributing factor to the diabetic prothrombotic state is the aberrant regulation of antiplatelet-activating mechanisms that normally maintain high levels of inhibitory cAMP to prevent aggregation. Molecules directly affecting platelet cAMP production are the arachidonic acid metabolites TXA_2 and prostacyclin (PGI2). TXA_2 is produced in the platelets themselves and is a positive-feedback mediator of platelet activation, while PGI2 is produced in endothelial cells and is an inhibitor of platelet aggregation. The platelet activator TXA_2 is synthesized by cyclooxygenase 1 (COX-1) in the platelets; the inhibition of COX-1 (with aspirin), which is irreversible and semi-selective, could improve platelet reactivity of T1DM subjects [104].

18.3.2 Cardiovascular diseases

The numbers of platelet GP receptors are found to enhance in cardiovascular patients and diabetic subjects. The HbA1c is the useful marker to detect the diabetes in individuals. The CD40L on platelets has been found to be correlated with this HbA1c concentration, as an upregulation of the CD40-CD40L system has been observed in diabetes mellitus [56] along with an increase in GPIa/IIa, GPIIb/IIIa, P-selectin (CD62), vWF, and CD63 [105]. Patients with ischemic heart disease and depression concomitantly may have increased risk of thrombosis due to abnormal platelet activation. Moreover, high vWF may increase the chance of cardiovascular disease. Elevated plasma levels of vWF are associated with established cardiovascular risk factors such as age, smoking, cholesterol, diabetes mellitus, and hypertension [105, 106]. Moreover, raised levels of vWF are predictive of stroke and vascular events among patients with atrial fibrillation [106].

18.3.3 Neurodegeneration

Platelet dysfunctions associated with aging can be linked to molecular alterations affecting several cellular systems that include cytoskeleton rearrangements, signal transduction, vesicular trafficking, and protein degradation. Aging in platelets and their age-dependent dysfunctions are of interest when evaluating the contribution of aging to the onset of aging-dependent pathologies, such as those affecting the nervous system linked to neurodegenerative diseases including Alzheimer's disease (AD), Parkinson's disease (PD), multiple sclerosis (MS), Huntington's disease (HD), and amyotrophic lateral sclerosis (ALS) [107, 108].

It is well understood from the above information that the platelets have two important types of granules, dense granules and alpha granules. The dense granules are loaded with proaggregatory factors (e.g., serotonin, calcium, and ADP), and during platelet activation, these granules release their content to the open canalicular system to be expelled out by the platelet, whereas the alpha granules have many hemostatic proteins (e.g., platelet-derived growth factor, vWF, and fibrinogen) [108]. Aging-related variations in the expression of specific platelet receptors and platelet activators have been reported and exemplified by a decrease in the number of receptors for PGI2 potent inhibitor of platelet function and high levels of TXA2 activator of platelet function in older individuals [109]. Even though the most important function of platelets is to prevent bleeding, they also play an important function in pathological conditions including neurological and neurodegenerative diseases (e.g., PD, schizophrenia, and AD). It is also important that platelets show high expression of several proteins associated with the development of AD, such as the APP amyloid precursor protein (APP) and tau protein. Additionally, platelets express enzymes involved in protein modifications such as glycogen synthase kinase 3 β (GSK-3β), α-, β-, and γ-secretases. Platelets have been compared with neurons because they have many biochemical similarities [108], as they have the storage and release capacity of neurotransmitters from platelets such as serotonin, glutamate, and dopamine [110, 111] and the expression of neuron-related proteins such as NMDA receptors [112]. Together, this makes it interesting to consider the contribution of platelets to the hallmarks of neurodegeneration.

It is very imperative to note that the brains of cerebral amyloid angiopathy (CAA) patients (a disorder characterized by deposits of Aβ40 in cerebral arteries and capillaries) is estimated its prevalence in 90–98% of AD patients and this AD is present in 30% of individuals without dementia over 60 years old. It is needless to mention here that AD is related with the amyloid-beta (Aβ) pathogenesis. Aβ40 peptide activates and promotes platelet adhesion and aggregation [113] by different receptors such as CD36 and GPIbα, triggering several signal transduction pathways involving p38MAPK and COX1 and synthesis of TXA2, which ultimately increase Ca^{2+} levels, activates calpain, and increases Aβ40 peptide secretion [114]. The thrombin receptor PAR1 could also have a role in the consequent activation of p38 MAPK and cytosolic phospholipase A2 (PLA2), and TXA2 formation. Aβ40 peptides modify platelet shape change and granule release through activation of the small GTPase RhoA and phosphorylation of its downstream effector, myosin light chain kinase, involving cytoskeletal reorganization [113]. In platelets, the production of Aβ40 peptides also regulates the platelets phosphatidylserine exposure which has been found to be involved for further increase in platelet Aβ40 levels. A correlation between increased ROS formation in AD platelets and increased oxidative stress in AD patients has been demonstrated [108].

The coagulation cascade plays the critical role in the development of an inflammatory response in MS. Platelets are trapped in chronic active demyelinating MS lesion. The paralysis and experimental autoimmune encephalomyelitis were found to be ameliorated and reduced after inhibiting (the GPIIb/IIIa blocker, abciximab) the main platelets integrin GPIIb/IIIa [115]. A role of thrombin cascade in the development of inflammation in MS has also pointed out [116]. This multifunctional cell (platelets) is activated by different endogenous, physiological agonists, including ADP, collagen, or thrombin, due to the vast number of receptors present on the surface of platelets, as discussed previously. During the vessel wall injury, the circulating platelets are found to (a) immobilize immediately, (b) interact with the vWF binding to collagen and the glycoprotein GPIb-V-IX complex [117], and (c) initiate adhesion of free moving platelets of circulation to the subendothelial extracellular matrix. The information about involvement and changes of platelet receptors in other aging-induced neurodegenerative disorders is yet to be studied.

18.4 Cancer

The knowledge in the last decade on how tumor cells exploit platelets for survival, arrest, and finally extravasation from blood vessels to distant organs has tremendously increased [118]. Platelets protect circulating tumor cells (CTCs) by encasing tumor cells in a thrombus (**Figure 2**), protecting them from cytolysis by natural killer cells [119]. Tumor cells activate platelets for a stable adhesion between platelets and tumor cells by distinct mechanisms, which are the reasons for hypercoagulation and increased risks of thrombosis in cancer patients [107]. Tumor cells release soluble mediators such as ADP, TXA2, or high-mobility group box 1 (HMGB1), which ligates with Toll-like receptor 4 (TLR4) to instigate local platelet activation [120]. The heterogeneous cloaks of platelets and tumor cells protect tumor cells from high shear forces in the blood circulation and from attack by leukocytes by a pseudonormal phenotype, and allow tumor cells to downregulate MHC class I molecules to escape T cell immune surveillance [118]. This finally leads to tumor growth and angiogenesis with a deposition of platelets in tumors by the engagement of the P-selectin-αIIbβ3-talin complex [121]. This P-selectin binding to its ligands can also activate pro-survival kinases, resulting in enhanced tumor growth in mice neuroblastoma cells [122]. Platelets, those bind to colorectal cancer cells by means of P-selectin in the presence of polymorphonuclear leukocytes, were able to activate human microvascular endothelial cells. This in turn is found to express inflammatory proteins, e.g., chemokine CCL5 most abundantly [123]. CCL5 recruited monocytes to the metastatic microenvironment that finally culminated in an augmented number of metastatic foci in the lungs. It suggests that platelets and platelet-derived P-selectin seem to be crucially involved in cancer immunity (**Figure 2**).

After platelet activation, αIIbβ3 can switch to at least one of the two different active ligand-binding states; both differ in their affinity for fibrinogen, which indicates the contribution of αIIbβ3 to tumor cell platelet interaction and aggregation [118, 124]. Hence, integrin αIIbβ3 is for several reasons an attractive target in hematogenous cancer cell dissemination. Different kinds of drugs which efficiently inhibit integrin αIIbβ3 have been approved for reduction or prevention of thrombotic cardiovascular events. Furthermore, integrin αIIbβ3 is capable of mediating bidirectional signaling [118]. On the one hand, binding of integrin αIIbβ3 to tumor cells can finally culminate in platelet activation, while on the other hand, platelet activation by, e.g., ADP, TXA2, or thrombin, can transfer αIIbβ3 to an active binding state [107]. These inside-out signaling confers the ability to bind several ligands (e.g., melanoma cell expressed αvβ3), which can induce protumorigenic and proangiogenic signals [125].

In different types of cancers, for instance in squamous cell carcinoma, lung and skin cancer, mesotheliomas, and cancer-associated fibroblasts, an

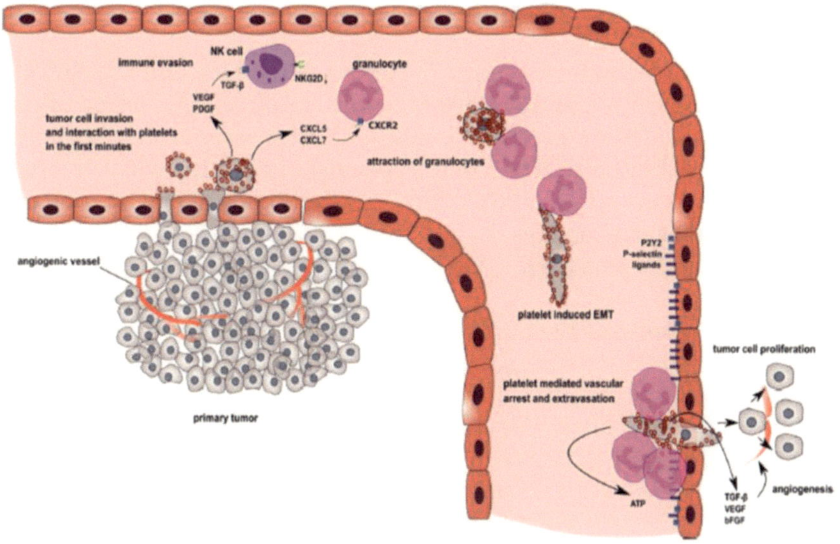

Figure 2.
Role of platelets in the metastatic cascade. Cells from the primary tumors are detached and invaded into the blood circulation. Platelets are activated immediately by their invasion and encasing the invaded tumor cells. These activated platelets are able to shift tumor cells to the vascular wall on another site of a distant organ and got arrested via interaction of P-selectin and PSGL-1 by facilitating tumor cell extravasation to the subendothelial matrix by endothelial P2Y2 receptor activation. (This diagram is adopted from Schlesinger [118].)

upregulated expression of podoplanin was detected [118, 126], converting an epithelial to mesenchymal transition (EMT) in MDCK cells, by increasing cell migration, and was associated with tumor invasion [127]. In healthy human tissues, podoplanin is expressed in lymphatic endothelial cells, osteocytes, keratinocytes, podocytes, and myofibroblasts among many other cell entities [128]. Podoplanin deficiency leads to form defective lymphatic vessels and causes death due to respiratory failure after birth [129]. Thus, the CLEC-2-podoplanin axis is an interesting target in course of hematogenous metastasis (especially for those patients with podoplanin-positive tumors), though more study is needed. Meanwhile, the tumor cell-induced platelets aggregating effect of the podoplanin CLEC-2 interplay has been revealed in several animal models, and by this way, different antibodies targeting different epitopes are generated [118, 130]. CLEC-2 ligand responsible for pronounced platelet aggregation and a podoplanin recognition domain in CLEC-2 was elucidated [131]. Podoplanin contains three platelet aggregation-stimulating domains in the extracellular section which is crucial for the aggregating function [132]. GPVI cytoplasmic tail is associated through a salt bridge with the Fc receptor γ chain (FcRγ), and upon ligand-mediated GPVI crosslinking and clustering, ITAM motif in FcRγ chains are unmasked and phosphorylated. This phosphorylated ITAM is sequentially found to recruit and activate Syk downstream signaling complex (composed of LAT and SLP76) to activate platelet and spreading of this platelet activation [133]. However, the study on the role of this ITAM-containing receptor in the interaction of platelets with tumor cells is very limited. Indeed, it has been observed that in C57BL/6J mice, deficiency in GPVI leads to decrease in thrombus formation with a prolongation of bleeding time [134]. Furthermore, an IgG-independent FcγRIIa-mediated cooperation between GPVI, GPIb-IX-V, and αIIbβ3 for platelet activation and spreading has been suggested. Studies dealing with the participation of FcγRIIa in tumor metastasis are barely available [135].

18.5 Platelet receptors and viral pathogens

There are different platelet receptors including TLRs, CRs, and DC-SIGN (as described in Sections 13, 16, and 17) that facilitate the direct interaction of platelets with viral pathogens. This interaction through different receptors causes both quantitative and qualitative dysfunctions of platelets associated with viral pathogens (**Table 2**). Platelets play a role in defending viral pathogen by binding of viral pathogen with platelets, which results not only in clearance of platelets but also clearance of viral pathogens [69].

18.5.1 Coronavirus disease-2019 (COVID-19)

Recent worldwide outbreak of COVID-19 is known to us as a flu-like disease, in which the respiratory illness (like the flu) with symptoms such as cough, fever, and in more severe cases, difficulty in breathing occurs. Early during its initiation (December, 2019) of infection, it was named as novel coronavirus disease-2019 or nCOVID-19. But in the next month (January, 2020), it was named as severe acute respiratory syndrome coronavirus 2 (SARS-CoV-2), which was later (February, 2020) designated as coronavirus disease 2019 or in short "COVID-19" by WHO (World Health Organization). Despite of its acute respiratory symptom, it is being reported that in more severe cases, abnormal clotting is a common phenomenon that culminates into the risk of cardiovascular disease (CVD), and this fact is the most crucial one here to discuss with (**Figure 3**). In severe cases of COVID-19 patients, clots in the small vessels of all organs, not only in the lungs but also in heart, liver, and kidney, have been found [136]. Though the involvement of any platelet receptor has not been found yet, the presence of clot indicates toward the involvement of platelet receptors for sure, as there are different platelet receptors already existing which can bind with different viruses (**Table 2**). Scientists have

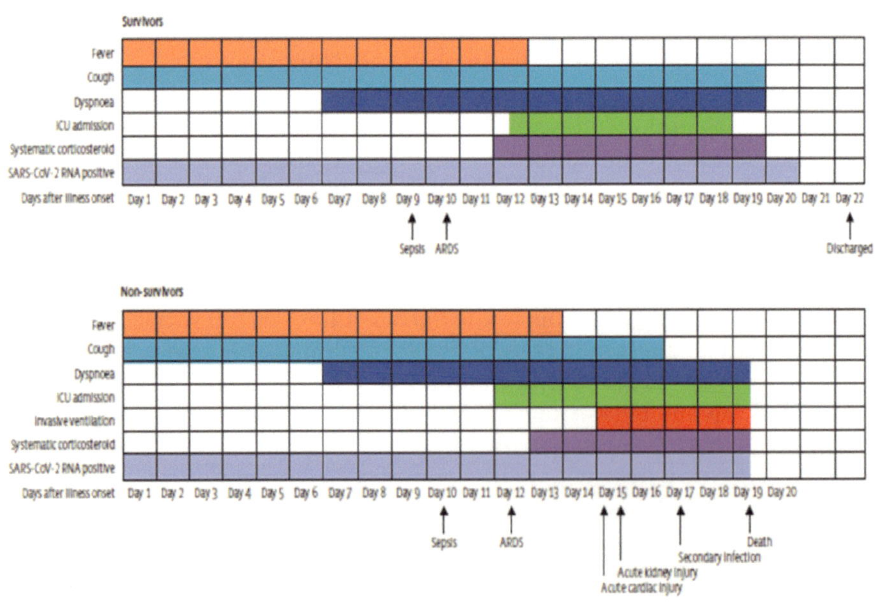

Figure 3.
Schematic presentations of clinical courses of major symptoms and outcomes and duration of viral shedding in the COVID-19 patients. ICU: intensive care unit; ARDS: acute respiratory distress syndrome; SARS-CoV-2: severe acute respiratory syndrome coronavirus 2; and COVID-19: coronavirus disease 2019. (This diagram is adopted from Zhou et al. [136].)

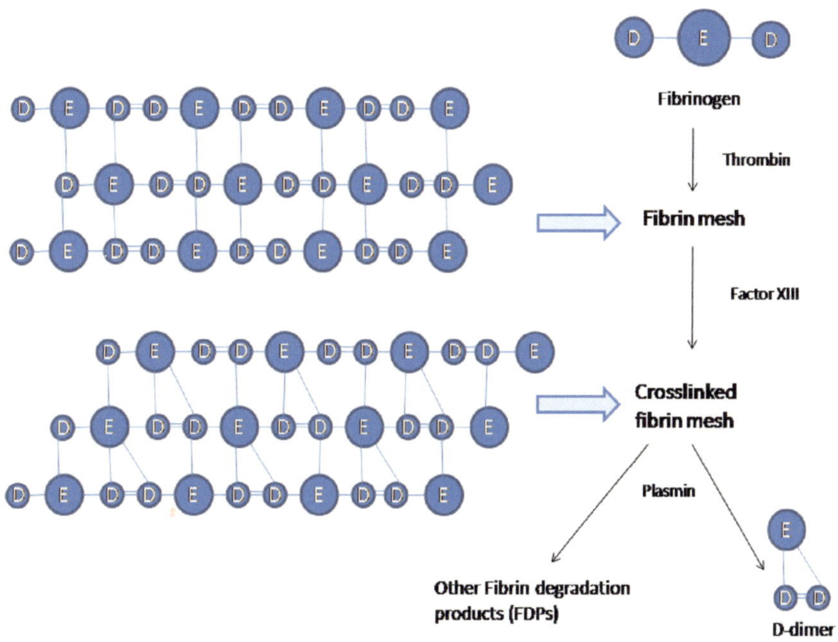

Figure 4.
Schematic diagram of D-dimer formation as a risk factor for the development of CVD in the COVID-19. Fibrinogen is made up of one E domain and two D domain, whereas D-dimer has the same number of domains but with different structural orientation with cross-linked D domain.

noticed till date about the presence of the biomarker of clot, known as D-dimer in the blood samples of COVID-19 patients with severity [136]. The presence of D-dimer in the circulation has its normal range of <0.5 μg/ml. D-dimers are not normally present in human blood plasma, except when the coagulation system has been activated (**Figure 4**). The structure of D-dimer is either a 180- or 195-kDa molecule of two D domains or a 340-kDa molecule of two D domains and one E domain of the original fibrinogen molecule [137]. D-dimer levels over 1 μg/ml at the time of admission predicted an almost 18-fold increase in odds of dying before discharge of nCOVID-19 patients seen at two hospitals in Wuhan, China [136]. D-dimer, a fibrin degradation product indicating thrombosis, can exceed 70 or 80 μg/ml, which clearly can indicate the severity of the illness. In that case, the anticoagulation therapy (with the anticoagulation, regardless of the underlying mechanism) may be initiated for severe COVID-19 patients, unless otherwise contraindicated the consequences [136].

19. Conclusion

Platelet receptors, particularly their adhesion receptors, execute an important role in the regulation of circulatory hemostasis. A rapid transition of circulating resting platelets to the activated state, adhesion, and aggregation in thrombus formation happen by the cascades of events. Briefly, these events are: (i) initial contact adhesion with the platelet GPIb-IX-V and GPVI, and collagen or collagen-bound vWF; (ii) activation, spreading, and secretion, involving GPIb-IX-V- and GPVI-dependent signaling pathways; (iii) secretion of agonists (e.g., ADP), the P2Y1/P2Y12 receptors activation, and upregulation of integrin αIIbβ3; and (iv) αIIbβ3-dependent aggregation, involving vWF or fibrinogen (**Table 1** and **Figure 1**). These processes happen very fast, and to imagine their rate of reactions, it may be informed that the initiation of thrombus formation starts within seconds after an injury, and complete thrombus is formed within minutes. Consequent to the aggregation process by recruiting free flowing inactivated platelets, the coagulation process accelerates on the activated platelet surface with the involvement of leukocytes and RBCs (red clot), followed by stabilization of the thrombus with the polymerized fibrin, and finally the αIIbβ3-mediated clot contraction *in vivo*. The dysfunction of GPIb-IX-V in humans (Bernard-Soulier syndrome) or mice (GPIbα gene knockouts) thrombus formation under high shear conditions is specifically impaired [5, 6]. Recent evidence also reveals a requirement for GPVI in stable thrombus formation at high shear rates, which provides the evidence of functional co-association of GPVI and GPIb-IX-V [6, 25]. The recent discovery of 5-HT receptors (2A and 3) and their role on platelet aggregation added a special attention on the interaction of neurotransmitter with the platelet adhesion receptors (**Table 1**), which may open a new avenue in geriatrics research. New molecular insights into the role of primary platelet adhesion receptors in vascular biology would be beneficial in pathophysiology study. GPIb-IX-V and GPVI together form a multifunctional adhesive cluster of proteins on the platelet surface that control thrombus formation at high shear and are also central to other vascular processes including inflammation and atherogenesis. Other platelet receptors including GPs, integrins, GPCRs, and Ig superfamily also play a pivotal role to maintain normal physiology. The physiology and pathophysiology of different platelet surface receptors at the molecular level have been described in this chapter. The molecular pathophysiology of different platelet receptors under different pathological conditions has been described. The recent advancement of the knowledge of how platelet is being considered as a biomarker of neurodegeneration and how far it is reliable to consider it (platelet) with central nervous system (CNS) has been pointed out with its updated information. The molecular aspects of inflammation, cancer (**Figure 2**), and even viral pathogens (**Table 2**), especially COVID-19 pathology (**Figures 3** and **4**) in relation to platelet receptors, are also taken under consideration to present in this chapter. This updated information will enlighten and motivate researchers to advance and carry forward their research with platelet at the molecular level of its receptor in different pathophysiological conditions. Finally, it may be stated that all of these information and thoughts may help to find out different biomarkers and treatment for different diseases in the near future.

Acknowledgements

The authors are thankful to the Department of Pharmaceutical Technology, Jadavpur University, Kolkata; University Grants Commission (UGC), New Delhi; Science and Engineering Research Board (SERB), Govt. of India; and Indian Council of Medical Research (ICMR), New Delhi, India, for the present work.

Conflict of interest

There is no conflict of interest.

Author details

Mrinal K. Poddar[1*] and Soumyabrata Banerjee[2]

1 Department of Pharmaceutical Technology, Jadavpur University, Kolkata, India

2 Department of Psychology, Neuroscience Program, Central Michigan University, Mount Pleasant, MI, USA

*Address all correspondence to: mrinalkp@yahoo.com

References

[1] Gulliver G. Notes of some of the researches in anatomy, physiology, pathology, and botany of George Gulliver, F.R.S., 1880, University of Glasglow Library (archive.org), England. Carpenter WB, Principles of human physiology. In: Power H. editor. 9th ed. London: J. & A. Churchill; 1881. Available from: https://archive.org/details/b21482111

[2] Bizzozero G. Su di un nuovo elemento morfologico del sangue dei mammiferi e della sua importanza nella trombosi e nella coagulazione. Amsterdam, Netherlands: Osservatore Gazzetta delle Cliniche. 1881;**17**:785-787

[3] Michelson AD. Platelets. 3rd ed. Academic Press; 2013

[4] Hosseini E, Mohtashami M, Ghasemzadeh M. Down-regulation of platelet adhesion receptors is a controlling mechanism of thrombosis, while also affecting posttransfusion efficacy of stored platelets. Thrombosis Journal. 2019;**17**:20. DOI: 10.1186/s12959-019-0209-5

[5] Andrews RK, Gardiner EE, Shen Y, Whisstock JC, Berndt MC. Glycoprotein Ib-IX-V. The International Journal of Biochemistry & Cell Biology. 2003;**35**:1170-1174

[6] Andrews RK, Gardiner EE, Shen Y, Berndt MC. Platelet interactions in thrombosis. IUBMB Life. 2004;**56**:13-18

[7] Kauskot A, Hoylaerts MF. Platelet receptors. Handbook of Experimental Pharmacology. 2012;**210**:23-57

[8] Simon DI, Chen Z, Xu H, Li CQ, Dong J-F, McIntire LV, et al. Platelet glycoprotein Ibα is a counterreceptor for the leukocyte integrin Mac-1 (CD11b/CD18). The Journal of Experimental Medicine. 2000;**192**:193-204

[9] Romo GM, Dong J-F, Schade AJ, Gardiner EE, Kansas GS, Li CQ, et al. The glycoprotein Ib-IX-V complex is a platelet counterreceptor for P selectin. The Journal of Experimental Medicine. 1999;**190**:803-814

[10] Whisstock JC, Shen Y, Lopez JA, Andrews RK, Berndt MC. Molecular modeling of the seven tandem leucine-rich repeats within the ligand-binding region of platelet glycoprotein Iba. Thrombosis and Haemostasis. 2002;**87**:329-333

[11] Shen Y, Romo GM, Dong J-F, Schade A, McIntire LV, Kenny D, et al. Requirement of leucine-rich repeats of glycoprotein (GP) Iba for shear-dependent and static binding of von Willebrand factor to the platelet membrane GPIb-IX-V complex. Blood. 2000;**95**:903-910

[12] Dumas JJ, Kumar R, McDonagh T, Sullivan F, Stahl ML, Somers WS, et al. Crystal structure of the wildtype von Willebrand factor A1-glycoprotein Iba complex reveals conformation differences with a complex bearing von Willebrand disease mutations. The Journal of Biological Chemistry. 2004;**279**:23327-23334

[13] Ramakrishnan V, DeGuzman F, Bao M, Hall SW, Leung LL, Phillips DR. A thrombin receptor function for platelet glycoprotein Ib-IX unmasked by cleavage of glycoprotein V. Proceedings of the National Academy of Sciences of the United States of America. 2001;**98**:1823-1828

[14] Huizinga EG, Tsuji S, Romijn RA, Schiphorst ME, de Groot PG, Sixma JJ, et al. Structures of glycoprotein Iba and its complex with von Willebrand factor A1 domain. Science. 2002;**297**:1176-1179

[15] De Cristofaro R, De Candia E. Thrombin domains: Structure, function and interaction with platelet

receptors. Journal of Thrombosis and Thrombolysis. 2003;**15**:151-163

[16] Yun TH, Baglia FA, Myles T, Navaneetham D, Lopez JA, Walsh PN, et al. Thrombin activation of factor XI on activated platelets requires the interaction of factor XI and platelet glycoprotein Iba with thrombin anion-binding exosites I and II, respectively. The Journal of Biological Chemistry. 2003;**278**:48112-48119

[17] Bradford HN, Pixley RA, Colman RW. Human factor XII binding to the glycoprotein Ib-IX-V complex inhibits thrombin-induced platelet aggregation. The Journal of Biological Chemistry. 2000;**275**:22756-22763

[18] Baglia FA, Badellino KO, Li CQ, Lopez JA, Walsh PN. Factor XI binding to the platelet glycoprotein Ib-IX-V complex promotes factor XI activation by thrombin. The Journal of Biological Chemistry. 2002;**277**:1662-1668

[19] Adam F, Bouton MC, Huisse MG, Jandrot-Perrus M. Thrombin interaction with platelet membrane glycoprotein Ib alpha. Trends in Molecular Medicine. 2003;**9**:461-464

[20] De Marco L, Mazzucato M, Masotti A, Ruggeri ZM. Localization and characterization of an alpha-thrombin-binding site on platelet glycoprotein Ib alpha. The Journal of Biological Chemistry. 1994;**269**:6478-6484

[21] Lanza F. Bernard-Soulier syndrome (hemorrhagiparous thrombocytic dystrophy). Orphanet Journal of Rare Diseases. 2006;**1**:46

[22] Kroll MH, Hellums JD, McIntire LV, Schafer AI, Moake JL. Platelets and shear stress. Blood. 1996;**88**:1525-1541

[23] Smethurst PA, Joutsi-Korhonen L, O'Connor MN, Wilson E, Jennings NS, Garner SF, et al. Identification of the primary collagen-binding surface on human glycoprotein VI by site directed mutagenesis and by a blocking phage antibody. Blood. 2004;**103**:903-911

[24] Lecut C, Arocas V, Ulrichts H, Elbaz A, Villeval JL, Lacapere JJ, et al. Identification of residues within human GPVI involved in the binding to collagen: Evidence for the existence of distinct binding sites. The Journal of Biological Chemistry. 2004;**279**:52293-52299

[25] Gardiner EE, Arthur JF, Kahn ML, Berndt MC, Andrews RK. Regulation of platelet membrane levels of glycoprotein VI by a platelet-derived metalloproteinase. Blood. 2004;**104**:3611-3617

[26] Yip J, Shen Y, Berndt MC, Andrews RK. Primary platelet adhesion receptors. IUBMB Life. 2005;**57**:103-108

[27] Clemetson KJ, Clemetson JM. Platelet collagen receptors. Thrombosis and Haemostasis. 2001;**86**:189-197

[28] Miura Y, Takahashi T, Jung SM, Moroi M. Analysis of the interaction of platelet collagen receptor glycoprotein VI (GPVI) with collagen. A dimeric form of GPVI, but not the monomeric form, shows affinity to fibrous collagen. The Journal of Biological Chemistry. 2002;**277**:46197-46204

[29] Kasirer-Friede A, Kahn ML, Shattil SJ. Platelet integrins and immunoreceptors. Immunological Reviews. 2007;**218**:247-264

[30] Xiong JP, Stehle T, Goodman SL, Arnaout MA. New insights into the structural basis of integrin activation. Blood. 2003;**102**:1155-1159

[31] Salsmann A, Schaffner-Reckinger E, Kieffer N. RGD, the Rho'd to cell spreading. European Journal of Cell Biology. 2006;**85**:249-254

[32] Nurden AT. Glanzmann thrombasthenia. Orphanet Journal of Rare Diseases. 2006;**1**:10

[33] Ozaki Y, Suzuki-Inoue K, Inoue O. Novel interactions in platelet biology: CLEC-2/podoplanin and laminin/GPVI. Journal of Thrombosis and Haemostasis. 2009;7(Suppl 1):191-194

[34] Nuyttens BP, Thijs T, Deckmyn H, Broos K. Platelet adhesion to collagen. Thrombosis Research. 2011;**127** (Suppl 2):S26-S29

[35] Siljander PR, Hamaia S, Peachey AR, Slatter DA, Smethurst PA, Ouwehand WH, et al. Integrin activation state determines selectivity for novel recognition sites in fibrillar collagens. The Journal of Biological Chemistry. 2004;**279**:47763-47772

[36] Cosemans JM, Iserbyt BF, Deckmyn H, Heemskerk JW. Multiple ways to switch platelet integrins on and off. Journal of Thrombosis and Haemostasis. 2008;**6**:1253-1261

[37] Suzuki-Inoue K, Fuller GL, Garcia A, Eble JA, Pohlmann S, Inoue O, et al. A novel Syk-dependent mechanism of platelet activation by the C-type lectin receptor CLEC-2. Blood. 2006;**107**:542-549

[38] Patrono C, Garcia Rodriguez LA, Landolfi R, Baigent C. Low-dose aspirin for the prevention of atherothrombosis. The New England Journal of Medicine. 2005;**353**:2373-2383

[39] Habib A, FitzGerald GA, Maclouf J. Phosphorylation of the thromboxane receptor alpha, the predominant isoform expressed in human platelets. The Journal of Biological Chemistry. 1999;**274**:2645-2651

[40] Hirata T, Ushikubi F, Kakizuka A, Okuma M, Narumiya S. Two thromboxane A2 receptor isoforms in human platelets. Opposite coupling to adenylyl cyclase with different sensitivity to Arg60 to Leu mutation. The Journal of Clinical Investigation. 1996;**97**:949-956

[41] Paul BZ, Ashby B, Sheth SB. Distribution of prostaglandin IP and EP receptor subtypes and isoforms in platelets and human umbilical artery smooth muscle cells. British Journal of Haematology. 1998;**102**:1204-1211

[42] Ma H, Hara A, Xiao CY, Okada Y, Takahata O, Nakaya K, et al. Increased bleeding tendency and decreased susceptibility to thromboembolism in mice lacking the prostaglandin E receptor subtype EP(3). Circulation. 2001;**104**:1176-1180

[43] Gross S, Tilly P, Hentsch D, Vonesch JL, Fabre JE. Vascular wall-produced prostaglandin E2 exacerbates arterial thrombosis and atherothrombosis through platelet EP3 receptors. The Journal of Experimental Medicine. 2007;**204**:311-320

[44] Philipose S, Konya V, Sreckovic I, Marsche G, Lippe IT, Peskar BA, et al. The prostaglandin E2 receptor EP4 is expressed by human platelets and potently inhibits platelet aggregation and thrombus formation. Arteriosclerosis, Thrombosis, and Vascular Biology. 2010;**30**:2416-2423

[45] Stitham J, Arehart EJ, Gleim SR, Douville KL, Hwa J. Human prostacyclin receptor structure and function from naturally-occurring and synthetic mutations. Prostaglandins & Other Lipid Mediators. 2007;**82**:95-108

[46] Oestreich J. SCH-530348, a thrombin receptor (PAR-1) antagonist for the prevention and treatment of atherothrombosis. Current Opinion in Investigational Drugs. 2009;**10**:988-996

[47] Himanen JP. Ectodomain structures of Eph receptors. Seminars in Cell & Developmental Biology. 2012;**23**:35-42

[48] Prevost N, Woulfe D, Tognolini M, Brass LF. Contact-dependent signaling during the late events of platelet activation. Journal of Thrombosis and Haemostasis. 2003;**1**:1613-1627

[49] Gachet C. P2 receptors, platelet function and pharmacological implications. Thrombosis and Haemostasis. 2008;**99**:466-472

[50] Oury C, Sticker E, Cornelissen H, De Vos R, Vermylen J, Hoylaerts MF. ATP augments von Willebrand factor-dependent shear-induced platelet aggregation through Ca^{2+}-calmodulin and myosin light chain kinase activation. The Journal of Biological Chemistry. 2004;**279**:26266-26273

[51] Hu H, Hoylaerts MF. The P2X1 ion channel in platelet function. Platelets. 2010;**21**:153-166

[52] Angelillo-Scherrer A, Burnier L, Flores N, Savi P, DeMol M, Schaeffer P, et al. Role of Gas6 receptors in platelet signaling during thrombus stabilization and implications for antithrombotic therapy. The Journal of Clinical Investigation. 2005;**115**:237-246

[53] Cosemans JM, Van Kruchten R, Olieslagers S, Schurgers LJ, Verheyen FK, Munnix IC, et al. Potentiating role of Gas6 and Tyro3, Axl and Mer (TAM) receptors in human and murine platelet activation and thrombus stabilization. Journal of Thrombosis and Haemostasis. 2010;**8**:1797-1808

[54] Orlowski E, Chand R, Yip J, Wong C, Goschnick MW, Wright MD, et al. A platelet tetraspanin superfamily member, CD151, is required for regulation of thrombus growth and stability in vivo. Journal of Thrombosis and Haemostasis. 2009;**7**:2074-2084

[55] Goschnick MW, Lau LM, Wee JL, Liu YS, Hogarth PM, Robb LM, et al. Impaired "outside-in" integrin alphaIIbbeta3 signaling and thrombus stability in TSSC6-deficient mice. Blood. 2006;**108**:1911-1918

[56] Nergiz-Unal R, Rademakers T, Cosemans JM, Heemskerk JW. CD36 as a multiple-ligand signaling receptor in atherothrombosis. Cardiovascular & Hematological Agents in Medicinal Chemistry. 2011;**9**:42-55

[57] Roberts W, Magwenzi S, Aburima A, Naseem KM. Thrombospondin-1 induces platelet activation through CD36-dependent inhibition of the cAMP/protein kinase A signaling cascade. Blood. 2010;**116**:4297-4306

[58] Ghosh A, Murugesan G, Chen K, Zhang L, Wang Q, Febbraio M, et al. Platelet CD36 surface expression levels affect functional responses to oxidized LDL and are associated with inheritance of specific genetic polymorphisms. Blood. 2011;**117**:6355-6366

[59] Giomarelli B, Washington VA, Chisholm MM, Quigley L, McMahon JB, Mori T, et al. Inhibition of thrombin-induced platelet aggregation using human single-chain Fv antibodies specific for TREM-like transcript-1. Thrombosis and Haemostasis. 2007;**97**:955-963

[60] Washington AV, Gibot S, Acevedo I, Gattis J, Quigley L, Feltz R, et al. TREM-like transcript-1 protects against inflammation-associated hemorrhage by facilitating platelet aggregation in mice and humans. The Journal of Clinical Investigation. 2009;**119**:1489-1501

[61] Nanda N, Bao M, Lin H, Clauser K, Komuves L, Quertermous T, et al. Platelet endothelial aggregation receptor 1 (PEAR1), a novel epidermal growth factor repeat-containing transmembrane receptor, participates in platelet contact-induced activation. The Journal of Biological Chemistry. 2005;**280**:24680-24689

[62] Herrera-Galeano JE, Becker DM, Wilson AF, Yanek LR, Bray P, Vaidya D, et al. A novel variant in the platelet endothelial aggregation receptor-1 gene is associated with increased platelet aggregability. Arteriosclerosis, Thrombosis, and Vascular Biology. 2008;**28**:1484-1490

[63] Jones CI, Bray S, Garner SF, Stephens J, de Bono B, Angenent WG, et al. A functional genomics approach reveals novel quantitative trait loci associated with platelet signaling pathways. Blood. 2009a;**114**:1405-1416

[64] Johnson AD, Yanek LR, Chen MH, Faraday N, Larson MG, Tofler G, et al. Genome-wide meta-analyses identifies seven loci associated with platelet aggregation in response to agonists. Nature Genetics. 2010;**42**:608-613

[65] Morel O, Morel N, Freyssinet JM, Toti F. Platelet microparticles and vascular cells interactions: A checkpoint between the haemostatic and thrombotic responses. Platelets. 2008;**19**:9-23

[66] Jones CI, Garner SF, Moraes LA, Kaiser WJ, Rankin A, Ouwehand WH, et al. PECAM-1 expression and activity negatively regulate multiple platelet signaling pathways. FEBS Letters. 2009b;**583**:3618-3624

[67] Mori J, Pearce AC, Spalton JC, Grygielska B, Eble JA, Tomlinson MG, et al. G6b-B inhibits constitutive and agonist-induced signaling by glycoprotein VI and CLEC-2. The Journal of Biological Chemistry. 2008;**283**:35419-35427

[68] Freson K, Hashimoto H, Thys C, Wittevrongel C, Danloy S, Morita Y, et al. The pituitary adenylate cyclase activating polypeptide is a physiological inhibitor of platelet activation. The Journal of Clinical Investigation. 2004;**113**:905-912

[69] Seyoum M, Enawgaw B, Melku M. Human blood platelets and viruses: Defense mechanism and role in the removal of viral pathogens. Thrombosis Journal. 2018;**16**:16

[70] Beutler B. Inferences, questions and possibilities in toll-like receptor signalling. Nature. 2004;**430**:257-263

[71] Clark SR, Ma AC, Tavener SA, McDonald B, Goodarzi Z, Kelly MM, et al. Platelet TLR4 activates neutrophil extracellular traps to ensnare bacteria in septic blood. Nature Medicine. 2007;**13**:463-469

[72] Cloutiera N, Allaeysa I, Marcouxa G, Machlusb KR, Mailhotc B, Zuffereya A, et al. Platelets release pathogenic serotonin and return to circulation after immune complex-mediated sequestration. Proceedings of the National Academy of Sciences. 2018;**115**:E1550-E1559

[73] Herr N, Bode C, Duerschmied D. The effects of serotonin in immune cells. Frontiers in Cardiovascular Medicine. 2017;**4**:48

[74] Stratz C, Trenk D, Bhatia HS, Valina C, Neumann FJ, Fiebich BL. Identification of 5-HT3 receptors on human platelets: Increased surface immunoreactivity after activation with adenosine diphosphate (ADP) and thrombin receptor-activating peptide (TRAP). Thrombosis and Haemostasis. 2008;**99**:784-786

[75] Li N, Wallen NH, Ladjevardi M, Hjemdahl P. Effects of serotonin on platelet activation in whole blood. Blood Coagulation & Fibrinolysis. 1997;**8**(8):517-523

[76] Dale GL, Friese P, Batar P, Hamilton SF, Reed GL, Jackson KW, et al. Stimulated platelets use serotonin to enhance their retention of procoagulant proteins on the cell surface. Nature. 2002;**415**(6868):175-179

[77] Martel C, Cointe S, Maurice P, Matar S, Ghitescu M, Théroux P, et al.

Requirements for membrane attack complex formation and anaphylatoxins binding to collagen-activated platelets. PLoS One. 2011;**6**:e18812

[78] Chaipan C, Soilleux EJ, Simpson P, Hofmann H, Gramberg T, Marzi A, et al. DC-SIGN and CLEC-2 mediate human immunodeficiency virus type 1 capture by platelets. Journal of Virology. 2006;**80**:8951-8960

[79] Rivera J, Lozano ML, Navarro-Núñez L, Vicente V. Platelet receptors and signaling in the dynamics of thrombus formation. Haematologica. 2009;**94**:700-711. DOI: 10.3324/haematol.2008.003178

[80] Kulkarni S, Dopheide SM, Yap CL, Ravanat C, Freund M, Mangin P, et al. A revised model of platelet aggregation. The Journal of Clinical Investigation. 2000;**105**(6):783-791

[81] Ruggeri ZM. Factor v W. The Journal of Clinical Investigation. 1997;**100**(Suppl 11):S41-S46

[82] Hosseini E, Ghasemzadeh M. Different stages of platelet adhesion to the site of vascular injury. Iranian Journal of Blood and Cancer. 2012;**4**(3):133-142

[83] Mehran G. Investigation of Signaling Cross-talk Between Platelets and Neutrophils. Melbourne: Monash University; 2009

[84] Gibbins JM. Platelet adhesion signalling and the regulation of thrombus formation. Journal of Cell Science. 2004;**117**(16):3415-3425

[85] Nieswandt B, Watson SP. Platelet-collagen interaction: Is GPVI the central receptor? Blood. 2003;**102**:449-461

[86] Hosseini E, Ghasemzadeh M, Nassaji F, Jamaat ZP. GPVI modulation during platelet activation and storage: Its expression levels and ectodomain shedding compared to markers of platelet storage lesion. Platelets. 2017;**28**(5):498-508

[87] Gardiner EE, Andrews RK. Platelet receptor expression and shedding: Glycoprotein Ib-IX-V and glycoprotein VI. Transfusion Medicine Reviews. 2014;**28**(2):56-60

[88] Pennings GJ, Yong AS, Wong C, Al-Tamimi M, Gardiner EE, Andrews RK, et al. Circulating levels of soluble EMMPRIN (CD147) correlate with levels of soluble glycoprotein VI in human plasma. Platelets. 2014;**25**(8):639-642

[89] Goto S, Tamura N, Handa S, Arai M, Kodama K, Takayama H. Involvement of glycoprotein VI in platelet thrombus formation on both collagen and von Willebrand factor surfaces under flow conditions. Circulation. 2002;**106**:266-272

[90] Siljander PR, Munnix IC, Smethurst PA, Deckmyn H, Lindhout T, Ouwehand WH, et al. Platelet receptor interplay regulates collagen-induced thrombus formation in flowing human blood. Blood. 2004;**103**:1333-1341

[91] Gruner S, Prostredna M, Schulte V, Krieg T, Eckes B, Brakebusch C, et al. Multiple integrin-ligand interactions synergize in shear-resistant platelet adhesion at sites of arterial injury in vivo. Blood. 2003;**102**:4021-4027

[92] Massberg S, Gawaz M, Gruner S, Schulte V, Konrad I, Zohlnhofer D, et al. A crucial role of glycoprotein VI for platelet recruitment to the injured arterial wall in vivo. The Journal of Experimental Medicine. 2003;**197**:41-49

[93] Massberg S, Brand K, Gruner S, Page S, Muller E, Muller I, et al. A critical role of platelet adhesion in the initiation of atherosclerotic lesion

formation. The Journal of Experimental Medicine. 2002;**196**:887-896

[94] Weyrich AS, Lindemann S, Zimmerman GA. The evolving role of platelets in inflammation. Journal of Thrombosis and Haemostasis. 2003;**1**:1897-1905

[95] Wagner DD, Burger PC. Platelets in inflammation and thrombosis. Arteriosclerosis, Thrombosis, and Vascular Biology. 2003;**23**:2131-2137

[96] Bombeli T, Schwartz BR, Harlan JM. Adhesion of activated platelets to endothelial cells: Evidence for a GPIIb-IIIa dependent bridging mechanism and novel roles for endothelial intercellular adhesion molecule 1 (ICAM-1), avb3 integrin, and GPIba. The Journal of Experimental Medicine. 1998;**187**:329-339

[97] Bernard J, Soulier JP. Sur une nouvelle variété de dystrophie thrombocytaire-hémorragipare congénitale. La Semaine des Hôpitaux de Paris. 1948;**24**:3217-3222

[98] Poujol C, Ware J, Nieswandt B, Nurden AT, Nurden P. Absence of GPIb alpha is responsible for aberrant membrane development during megakaryocyte maturation: Ultrastructural study using a transgenic model. Experimental Hematology. 2002;**30**:352-360

[99] Pincus MR, Dykes DC, Carty RP, Miller JL. Conformational energy analysis of the substitution of Val for Gly 233 in a functional region of platelet GPIb alpha in platelet-type von Willebrand disease. Biochimica et Biophysica Acta. 1991;**1097**:133-139

[100] Franchini M, Montagnana M, Lippi G. Clinical, laboratory and therapeutic aspects of platelet-type von Willebrand disease. International Journal of Laboratory Hematology. 2008;**30**:91-94

[101] Alberio L, Friese P, Clemetson KJ, Dale GL. Collagen response and glycoprotein VI function decline progressively as canine platelets age *in vivo*. Thrombosis and Haemostasis. 2002;**88**:510-516

[102] Thattaliyath B, Cykowski M, Jagadeeswaran P. Young thrombocytes initiate the formation of arterial thrombi in zebrafish. Blood. 2005;**106**:118-124

[103] Bergmeier W, Piffath CL, Cheng G, Dole VS, Zhang Y, von Andrian UH, et al. Tumor necrosis factor-α-converting enzyme (ADAM17) mediates GPIbα shedding from platelets *in vitro* and *in vivo*. Circulation Research. 2004;**95**:677-683

[104] Wisinski JA, Kimple ME. Platelet dysfunction in type 1 diabetes: Stressing the thromboxanes. Diabetes. 2016;**65**:349-351

[105] Eibl N, Krugluger W, Streit G, Schrattbauer K, Hopmeier P, Schernthaner G. Improved metabolic control decreases platelet activation markers in patients with type-2 diabetes. European Journal of Clinical Investigation. 2004;**34**:205-209

[106] Spiel AO, Gilbert JC, Jilma B. Von Willebrand factor in cardiovascular disease focus on acute coronary syndromes. Circulation. 2008;**117**:1449-1459

[107] Banerjee S, Poddar MK. Aging and mammalian platelet biomarker. Thrombosis & Haemostasis: Research. 2019;**3**:1034

[108] Espinosa-Parrilla Y, Gonzalez-Billault C, Fuentes E, Palomo I, Alarcón M. Decoding the role of platelets and related microRNAs in aging and neurodegenerative disorders. Frontiers in Aging Neuroscience. 2019;**11**:151

[109] Simon LM, Edelstein LC, Nagalla S, Woodley AB, Chen ES, Kong X, et al. Human platelet microRNA-mRNA networks associated with age and gender revealed by integrated plateletomics. Blood. 2014;**123**:e37-e45

[110] Cupello A, Favale E, Audenino D, Scarrone S, Gastaldi S, Albano C. Decrease of serotonin transporters in blood platelets after epileptic seizures. Neurochemical Research. 2005;**30**:425-428

[111] Rainesalo S, Keranen T, Saransaari P, Honkaniemi J. GABA and glutamate transporters are expressed in human platelets. Brain Research. Molecular Brain Research. 2005;**141**:161-165

[112] Kalev-Zylinska ML, Green TN, Morel-Kopp MC, Sun PP, Park YE, Lasham A, et al. N-methyl-D-aspartate receptors amplify activation and aggregation of human platelets. Thrombosis Research. 2014;**133**: 837-847. DOI: 10.1016/j. thromres. 2014.02.011

[113] Canobbio I, Catricalà S, Di Pasqua LG, Guidetti G, Consonni A, Manganaro D, et al. Immobilized amyloid Aβ peptides support platelet adhesion and activation. FEBS Letters. 2013;**587**:2606-2611. DOI: 10.1016/j. febslet.2013.06.041

[114] Herczenik E, Bouma B, Korporaal SJ, Strangi R, Zeng Q, Gros P, et al. Activation of human platelets by misfolded proteins. Arteriosclerosis, Thrombosis, and Vascular Biology. 2007;**27**:1657-1665

[115] Langer HF, Choi EY, Zhou H, et al. Platelets contribute to the pathogenesis of experimental autoimmune encephalomyelitis. Circulation Research. 2012;**110**:1202-1210

[116] Satoh J. Molecular network analysis of multiple sclerosis brain lesion proteome. Nihon Rinsho Meneki Gakkai Kaishi. 2010;**4**:182-188

[117] Morel A, Bijak MB, Miller E, Rywaniak J, Miller S, Saluk J. Relationship between the increased haemostatic properties of blood platelets and oxidative stress level in multiple sclerosis patients with the secondary progressive stage. Oxidative Medicine and Cellular Longevity. 2015;**2015**:240918. 10 pages. DOI: 10.1155/2015/240918

[118] Schlesinger M. Role of platelets and platelet receptors in cancer metastasis. Journal of Hematology & Oncology. 2018;**11**:125

[119] Nieswandt B, Hafner M, Echtenacher B, Männel DN. Lysis of tumor cells by natural killer cells in mice is impeded by platelets. Cancer Research. 1999;**59**:1295-1300

[120] Yu L-X, Yan L, Yang W, Wu F-Q, Ling Y, Chen S-Z, et al. Platelets promote tumour metastasis via interaction between TLR4 and tumour cell-released high-mobility group box1 protein. Nature Communications. 2014;**5**:5256

[121] Qi C, Wei B, Zhou W, Yang Y, Li B, Guo S, et al. P-selectin-mediated platelet adhesion promotes tumor growth. Oncotarget. 2015;**6**:6584-6596

[122] Nolo R, Herbrich S, Rao A, Zweidler-McKay P, Kannan S, Gopalakrishnan V. Targeting P-selectin blocks neuroblastoma growth. Oncotarget. 2017;**8**:86657-86670

[123] Läubli H, Spanaus K-S, Borsig L. Selectin-mediated activation of endothelial cells induces expression of CCL5 and promotes metastasis through recruitment of monocytes. Blood. 2009;**114**:4583-4591

[124] Bennett JS. Regulation of integrins in platelets. Biopolymers. 2015;**104**:323-333

[125] Nemeth JA, Nakada MT, Trikha M, Lang Z, Gordon MS, Jayson GC, et al. Alpha-v integrins as therapeutic targets in oncology. Cancer Investigation. 2007;**25**:632-646

[126] Takemoto A, Miyata K, Fujita N. Platelet-activating factor podoplanin: From discovery to drug development. Cancer Metastasis Reviews. 2017;**36**:225-234

[127] Martín-Villar E, Megías D, Castel S, Yurrita MM, Vilaró S, Quintanilla M. Podoplanin binds ERM proteins to activate RhoA and promote epithelial mesenchymal transition. Journal of Cell Science. 2006;**119**:4541-4553

[128] Dang Q, Liu J, Li J, Sun Y. Podoplanin: A novel regulator of tumor invasion and metastasis. Medical Oncology. 2014;**31**:24

[129] Schacht V, Ramirez MI, Hong Y-K, Hirakawa S, Feng D, Harvey N, et al. T1alpha/podoplanin deficiency disrupts normal lymphatic vasculature formation and causes lymphedema. The EMBO Journal. 2003;**22**:3546-3556

[130] Nakazawa Y, Takagi S, Sato S, Oh-hara T, Koike S, Takami M, et al. Prevention of hematogenous metastasis by neutralizing mice and its chimeric anti-Aggrus/podoplanin antibodies. Cancer Science. 2011;**102**:2051-2057

[131] Kato Y, Kaneko MK, Kunita A, Ito H, Kameyama A, Ogasawara S, et al. Molecular analysis of the pathophysiological binding of the platelet aggregation-inducing factor podoplanin to the C-type lectin-like receptor CLEC-2. Cancer Science. 2008;**99**:54-61

[132] Toyoshima M, Nakajima M, Yamori T, Tsuruo T. Purification and characterization of the platelet-aggregating sialoglycoprotein gp44 expressed by highly metastatic variant cells of mouse colon adenocarcinoma 26. Cancer Research. 1995;**55**:767-773

[133] Hughes CE, Auger JM, McGlade J, Eble JA, Pearce AC, Watson SP. Differential roles for the adapters gads and LAT in platelet activation by GPVI and CLEC-2. Journal of Thrombosis and Haemostasis. 2008;**6**:2152-2159

[134] Konstantinides S, Ware J, Marchese P, Almus-Jacobs F, Loskutoff DJ, Ruggeri ZM. Distinct antithrombotic consequences of platelet glycoprotein Ib alpha and VI deficiency in a mouse model of arterial thrombosis. Journal of Thrombosis and Haemostasis. 2006;**4**:2014-2021

[135] Arman M, Krauel K. Human platelet IgG Fc receptor FcγRIIA in immunity and thrombosis. Journal of Thrombosis and Haemostasis. 2015;**13**:893-908

[136] Zhou F, Yu T, Du R, Fan G, Liu Y, Liu Z, et al. Clinical course and risk factors for mortality of adult inpatients with COVID-19 in Wuhan, China: A retrospective cohort study. Lancet. 2020;**395**:1054-1062

[137] Adam SS, Key NS, Greenberg CS. D-dimer antigen: Current concepts and future prospects. Blood. 2009;**113**:2878-2887

Chapter 3

Procoagulant Platelets

Andaleb Kholmukhamedov

Abstract

There are two well-known subpopulations of activated platelets: pro-aggregatory and procoagulant. Procoagulant platelets represent a subpopulation of activated platelets, which are morphologically and functionally distinct from pro-aggregatory ones. Although various names have been used to describe these platelets in the literature (CoaT, CoaTed, highly activated, ballooned, capped, etc.), there is a consensus on their phenotypic features including exposure of high levels of phosphatidylserine (PSer) on the surface; decreased aggregatory and adhesive properties; support of active tenase and prothrombinase complexes; maximal generation by co-stimulation of glycoprotein VI (GPVI) and protease-activated receptors (PAR). In this chapter, morphologic and functional features of procoagulant platelets, as well as the mechanisms of their formation, will be discussed.

Keywords: procoagulant, platelet, mitochondria, necrosis, apoptosis, coagulation

1. Introduction

Blood has different components, like plasma, red blood cells (RBC), white blood cells (WBC), and platelets. Platelets, although being only "tiny fragments of megakaryocytes (mother cell)," are essential for life. We need them, together with about two dozen of coagulation factors, to keep all that nutrient-rich liquid plasma, infection-fighting WBCs, and oxygen-carrying RBCs in our bodies in the case of trauma and bleeding. Platelets and coagulation get activated immediately upon being exposed to things they normally do not have contact with (e.g., during blood vessel wall rupture, collagen gets exposed and activates platelets). Upon activation, platelets form a mesh-like structure using another plasma protein called fibrinogen as bridges between them. This process forms a so-called "unstable clot." At later stages of blood clotting, generated thrombin converts soluble fibrinogen to insoluble fibrin, stabilizing the initial platelet plug. These processes are collectively called hemostasis. Every aspect of hemostasis has its history. Although exploring history can sometimes be tedious, studying the bridge between past and present is essential in basic understanding of the subject, the subject of procoagulant platelets in this case.

Unlike WBCs, which exist as functionally and morphologically distinct subpopulations, it has been thought for years that platelets are rather simple in their function, being just "cell particles." It was later revealed that platelets, although lacking nuclear material, are indeed very complex not only in their nature but also in function. Nowadays, the existence of two different subpopulations of activated platelets, pro-aggregatory and procoagulant, is a widely accepted fact. Pro-aggregatory platelets, historically known as activated, have been a major focus since the initiation of platelet research. The history of procoagulant platelets, on the other hand, can be subdivided into two periods. The first one is the discovery of platelet procoagulant functionality, whereas in the second period, procoagulant platelets were discovered and characterized as a distinct subpopulation of activated platelets.

In 1912, long before we learned that prothrombinase supporting platelets are distinct from pro-aggregatory ones, Howell discovered that unsaturated cephalin is a phospholipid factor that triggers clotting [1]. This discovery was followed by decades of controversial results of what phospholipid or what mixture of phospholipids is responsible for this effect until, in 1960, Karl Slotta demonstrated the presence of phosphatidylserine (PSer) is an absolute requirement [2]. This discovery established the central role of PSer in prothrombinase activity, which in a way, paused research activities in the procoagulant platelet field.

The history was resumed when two decades later, Bevers and colleagues spectrophotometrically showed that thrombin and collagen co-stimulated platelets possessing about 5-fold higher prothrombinase activity than those stimulated with thrombin or collagen alone. This effect was even more pronounced in the presence of exogenous factor Va [3]. Later that year, the same group confirmed their spectrophotometric observations in one-stage prothrombinase assay [4]. In 1985, Rosing et al. more thoroughly described the role of platelet PSer exposure in prothrombin (FII) and factor X (FX) activation [5]. They revealed that collagen and thrombin co-stimulation significantly increases prothrombinase and tenase activity, while this increase was abolished in the presence of phospholipase A_2 (PLA_2) from *N. naja*. Also observing no cellular lysis in these conditions, they concluded that PSer exposure, which is followed by thrombin and collagen co-stimulation, is responsible for this phenomenon. In 1993, with the discoveries of annexin V and advancement in flow cytometry, Dachary-Prigent et al. established a protocol to detect PSer exposing platelets, the methodology that has overtime become fundamental in procoagulant platelet field and has been widely used since then [6]. In 1997 Heemskerk and colleagues discovered that a percentage of platelets adhering to collagen, but not fibrinogen, balloon and expose PSer [7]. In parallel to these findings, there was a series of publications determining the extent and ultrastructure of PSer exposure, as well as the essential role of calcium in this process [8–10].

Although numerous studies described platelet procoagulant function in response to dual stimulation [3, 5, 11, 12], a breakthrough discovery that identified procoagulant platelets as a distinct subpopulation of activated platelets is the work by Alberio et al., where they demonstrated that only a certain percentage of activated platelets retain factor Va (FVa) on their surface [13]. This subpopulation was generated by co-stimulation with convulxin/collagen (GPVI agonists) and thrombin; hence they were named CoaT platelets. A few years down the road, Kulkarni and Jackson introduced a new term—'sustained calcium-induced platelet morphology (SCIP)' by discovering the fact that procoagulant platelets require prolonged elevation in cytosolic calcium to form [14]. Diversifying terminology of procoagulant platelets did not stop there, in 2005 revealing that procoagulant platelets retain multiple α-granule proteins on their surface in transglutaminase-dependent manner led to the introduction of a new term—'coated' (by α-granule proteins) platelets [15, 16]. Further studies by Panteleev et al. revealed that this subpopulation of activated platelets binds high levels of factors IXa and Xa [17]. In 2008, Jobe et al., characterizing molecular mechanisms of procoagulant platelet formation, introduced another term—highly activated platelets [18]. There are few more terms used in the literature like ballooned [7], ballooned and procoagulant-spread [19], and super-activated platelets [20].

Just like in a famous Indian parable, where blind men try to describe an elephant they have never encountered before by touching different parts of it, researchers have been describing different (morphological and functional) features of procoagulant platelets and introducing different terms based on their discoveries. Whereas indeed, everyone has been describing "different parts of the same elephant."

2. Mechanisms of procoagulant platelet formation

The discovery of procoagulant platelets as a distinct subpopulation of activated platelets at the beginning of the twenty-first century triggered research activities into cellular and molecular mechanisms of their formation. As shown by Alberio et al. and confirmed in later studies, procoagulant platelets are maximally generated upon co-stimulation of glycoprotein VI and PAR1/4. In 2005, Jobe and colleagues discovered that the absence of FcRγ, a key component responsible for glycoprotein VI signaling, ablates procoagulant platelet formation almost to baseline levels, evidencing GPVI stimulation is the major component of their generation [21]. The same year, Remenyi et al. demonstrated the role of mitochondrial permeability transition (MPT) in procoagulant platelet formation [16]. MPT is a Ca^{2+}-dependent molecular process that leads to mitochondrial swelling and cell death [22, 23]. During the onset of MPT, large pores are formed on the mitochondrial inner membrane making it non-specifically permeable to all solutes and molecules of molecular weight up to 1500 Da [24, 25]. It is a very well-known fact that mitochondrial Ca^{2+} overload can induce MPT, although the structure of MPT pore remains unknown.

Ca^{2+}, being a key signaling molecule in most cells, is important for many processes including platelet shape change and integrin $\alpha 2b\beta 3$ activation [26, 27]. In resting platelets, free Ca^{2+} is tightly regulated and maintained at about 100 nM in both the cytosol and mitochondria through the action of Ca^{2+}-ATPases in the plasma and cell membranes. Thus, platelet cytosolic-free Ca^{2+} is substantially lower than the blood Ca^{2+} levels, which are around 2 mM. With stimulation, however, cytosolic Ca^{2+} increases instantaneously. As outlined in **Figure 1**, this increase is mediated by activated phospholipase C (PLC). There are two major isoforms of PLC in human platelets, PLCβ and PLCγ. PLCβ is only activated downstream of Gq protein-coupled receptors (GPCRs). Whereas, PLCγ is activated downstream of numerous receptors like GPVI, glycoprotein Ib-IX complex (GPIb-IX), Fcγ receptor IIa (FcγRIIa), and C-type lectin-like receptor 2 (CLEC-2). As illustrated in **Figure 1**, both isoforms of PLC induce the release of Ca^{2+} from the dense tubular system (DTS) as well as activating transient receptor potential channel 3 (TRPC3). DTS release of Ca^{2+}, in turn, triggers its extracellular entry through store-operated calcium entry (SOCE). For some time, it remained mysterious on how the release of DTS calcium stores into the cytosol induces more Ca^{2+} to flow into the cell, further increasing its cytosolic concentrations. However, with the discovery of core components of SOCE, everything falls into place. STIM1 and ORAI1 are parts of the same complex. When STIM1 senses drop in Ca^{2+} levels within the DTS, it signals to ORAI1 located on a cell membrane to allow Ca^{2+} passage from extracellular space into the cell. As evidenced by the study in the early 2000s, increased cytosolic Ca^{2+} levels are one of the requirements of procoagulant formation [14]. Mitochondria, being the major and perhaps the only Ca^{2+} buffering system within platelets, equilibrate increased cytosolic Ca^{2+} following PAR and GPVI co-stimulation. The ensuing fate of mitochondrial Ca^{2+} overload is the opening of the mitochondrial permeability transition pore (MPTP), followed by PSer exposure and integrin deactivation (**Figure 1**), ultimately leading to a cell death. However, in this case cell death shall be considered physiologic considering the essential role of procoagulant platelets in hemostasis.

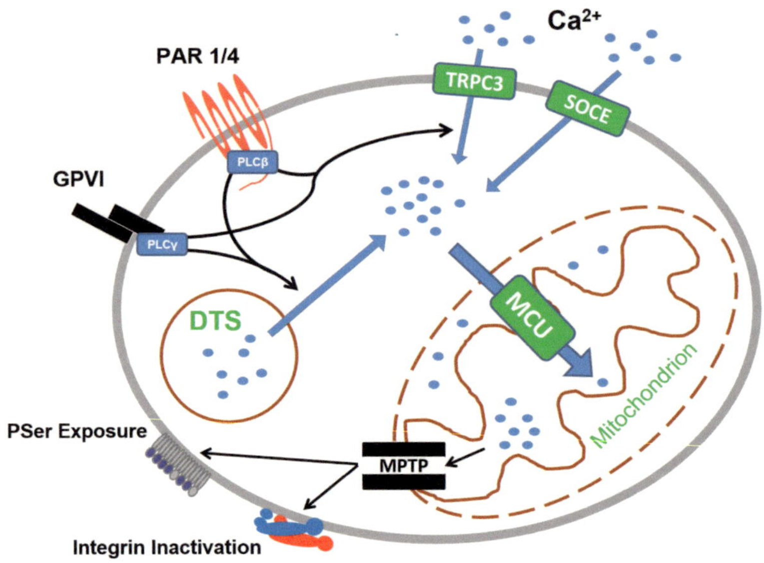

Figure 1.
Molecular mechanisms of physiologic agonist-induced procoagulant platelet formation. In resting platelets, low cytosolic Ca^{2+} is maintained by cell and plasma membrane Ca^{2+}-ATP-ases. Co-stimulation of protease-activated receptor 1/4 and glycoprotein VI leads to a sustained increase of cytosolic Ca^{2+} through activation of PLCβ and PLCγ, respectively. Increased cytosolic Ca^{2+} drives mitochondrial Ca^{2+} entry through the mitochondrial calcium uniporter complex. The resulting increase in mitochondrial Ca^{2+} opens mitochondrial permeability transition pore, which in turn leads to necrotic cell death and exposure of phosphatidylserine accompanied by integrin inactivation. DTS, dense tubular system; GPVI, glycoprotein VI; MCU, mitochondrial calcium uniporter; MPTP, mitochondrial permeability transition pore; PAR, protease-activated receptor; PSer, phosphatidylserine; SOCE, store-operated calcium entry; TRPC3, transient receptor potential channel 3. Adapted from [28] with modifications.

It brings us to another aspect of procoagulant platelet research, debated since their discovery, which is whether procoagulant platelets are necrotic or apoptotic. Cell death, both necrosis and apoptosis, is an essential event in the normal life of many cells in the human body, including platelets. Apoptosis, or programmed cell death, occurs in many organs throughout human lifetime. Necrosis, although considered to be mostly a catastrophic uncontrolled cell death, can also occur physiologically, as in the shedding of decidual endometrium during human menses. Necrosis and apoptosis differ in many aspects. The ultimate event during necrotic cell death is the osmotic swelling of the cell followed by the rupture of a cell membrane, whereas in apoptosis, cell shrinkage with preserved cell membrane is evident in later stages of this process.

Although the proposal that PSer is not homogeneously distributed on the surface of activated platelets was made back in 1985 [5], it took three decades to visualize that experimentally due to the complex nature of procoagulant platelets. In 2016, Podoplelova et al. presented a detailed structure of procoagulant platelets and cellular changes that give them the characteristic morphology. They elegantly demonstrated how the balloon is blown out from the platelet, leaving it as a "cap" [29]. The existence of bulges in a phospholipid bilayer, known as the open canalicular system (OCS), is essential, as it provides platelet a fair amount of surface reserve for ballooning. Morphologic appearance, in this case, resembles the classical osmotic swelling of necrotic cell death, as schematically presented on the left side of **Figure 2**. Another feature of necrosis, which is a collapse of energy production, is present in procoagulant platelets, as they lack energy-requiring contractile function [18]. Moreover, depletion of oxidative

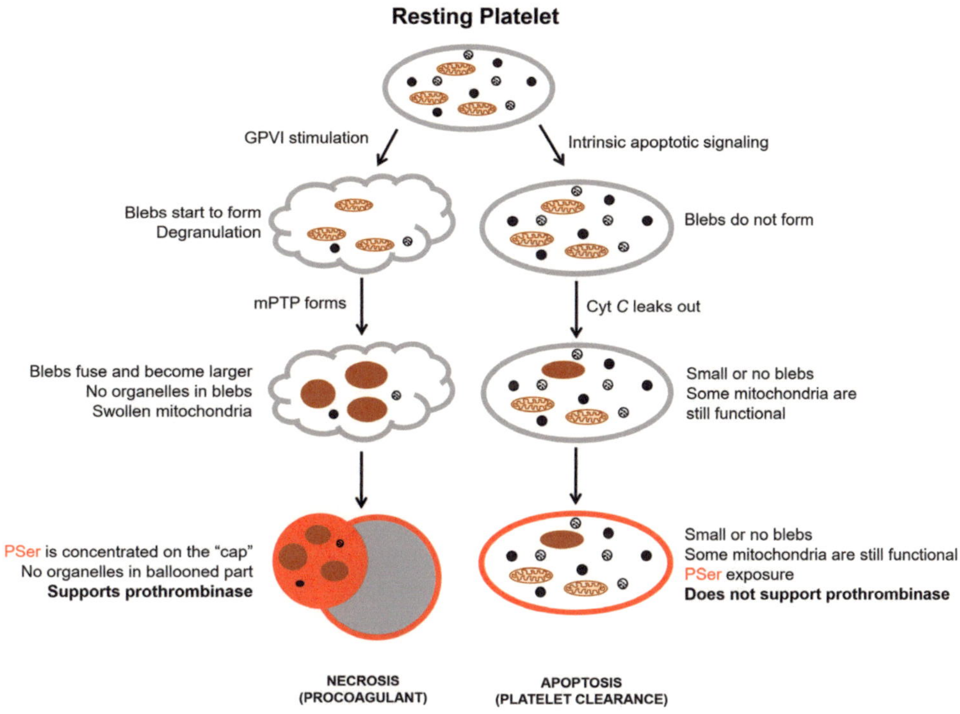

Figure 2.
Cellular mechanisms of platelet cell death. Cyt C, cytochrome C; GPVI, glycoprotein VI; mPTP, mitochondrial permeability transition pore; PSer, phosphatidylserine. PSer exposure is shown in red.

phosphorylation due to MPT, deactivates flippases that normally maintain the asymmetry of phospholipids. After the onset of MPT, contents of mitochondrial matrix get released to the cytosol leading to the second wave of cytosolic Ca^{2+} increase. This in turn leads to scramblase activation. Synergistically with deactivated flippases, scramblase performs its raison d'etre, which is flipping negatively charged PSer out in order to equilibrate its concentration. And finally, recent work demonstrated that platelets undergoing cell death through necrotic pattern and not apoptotic are functionally procoagulant as measured by prothrombinase support [30].

Platelet apoptosis (**Figure 2**, right side), on the other hand, has been widely implicated in platelet lifespan via the action of the intrinsic mitochondrial apoptosis pathway [31, 32]. Normally, antiapoptotic members of the BCL-2 family (e.g., BCL-xL) restrain the activity of proapoptotic Bax and Bak proteins, which are present within the cytosolic fraction of a platelet. When BCL-xL wears off, oligomerized Bax and Bak translocate to the mitochondrial outer membrane and permeabilize it, which leads to cytochrome *C* leakage into the cytosol, triggering apoptosome formation and eventually cell death. This is supported by the fact that the genetic ablation of murine BCL-xL reduces platelet life span from about 5 days to 5 hours [33, 34].

Thus, procoagulant platelets are indeed necrotic, while apoptosis is essential in platelet clearance. It should, however, be mentioned that the presence of PSer on the surface of a platelet is a signal for the reticuloendothelial system in the spleen and liver for clearance. Therefore, any procoagulant platelet that happened to escape the site of the active hemostatic process will be cleared out of the system by liver or spleen, just like aged apoptotic platelets do.

3. Functions of procoagulant platelets

The physiological relevance of procoagulant platelets had been questioned for a long time. In recent years, however, after the demonstration of a procoagulant platelet being predictive of bleeding or ischemic complications in patients with coronary artery disease, brain hemorrhage, traumatic brain injury, stroke, etc. [35–42] it is gaining more and more recognition. The importance of this subpopulation is further highlighted on a novel *ex vivo* model (which integrates all the core components of hemostasis [43]), where pharmacologic inhibition of platelet transition to a procoagulant state without affecting pro-aggregatory phenotype results in a decreased thrombus stability [44].

It was initially thought that the only function of procoagulant platelets is to support coagulation. However, with recent advances in the field, we learn that depending on their localizations, procoagulant platelets can play different functions within the thrombus. For general consideration, these two functionalities will be discussed separately here.

3.1 Coagulation support

Coagulation, a cascade of serine protease enzymatic reactions, is achieved by cleaving fibrinogen to fibrin, which transforms blood from a liquid to a gel-like state. Although platelet contribution to coagulation has been known for decades, the exact role of platelet phospholipids has been a matter of major debate. Dependence of hemostasis on biological membranes is very extensive, ranging from subendothelial membranes triggering coagulation and platelet activation to procoagulant platelet surface assembling tenase and prothrombinase complexes. It is very well known that at least two coagulation reactions are highly dependent on phospholipid surface, namely the activation of factor X and prothrombin by intrinsic tenase and prothrombinase, respectively. Both intrinsic tenase and prothrombinase are composed of the serine protease (FIXa for tenase and FXa for prothrombinase) and its protein cofactor (FVIIIa for tenase and FVa for prothrombinase). It is important to know that although both proteases alone are capable of activating their substrates, the presence of cofactors profoundly amplifies the catalytic process for up to 10,000 fold.

The current understanding of this process, based on numerous studies including mathematical modeling, is that phospholipid surface increases the rate of reactions by increasing the local concentration of coagulation factors, and thus increasing the probability of their interaction [29, 45–51]. This increase in a local concentration of factors, essential for tenase and prothrombinase complexes, is accomplished by the interaction of negatively charged gamma-carboxyglutamic acid (GLA) residues of the coagulation factors with negatively charged phosphatidylserine on the surface of procoagulant platelets. It is supported by the fact that GLA domain-containing (also known as vitamin K-dependent) factors predominantly bind to a procoagulant subpopulation of activated platelets [17, 29, 30]. Both enzymatic factors for intrinsic tenase (FIX) and prothrombinase (FX) are GLA domain-containing proteins and bind to PSer in a calcium-dependent manner. Whereas cofactors (FVIII and FV) structurally are not GLA proteins and bind to a procoagulant surface by different mechanisms.

In the case of FVIII, it has been shown by Gilbert et al. that it is not specific to a procoagulant surface [52]. It was further confirmed by Podoplelova et al. in their efforts exploring procoagulant platelet characteristics, they demonstrated that both (pro-aggregatory and procoagulant) subpopulations of activated platelets bind FVIIIa [29]. The fact that PS increases the catalytic activity of the intrinsic tenase complex by about 1500-fold [53] can be explained by FIXa's specificity to PSer. As outlined in **Figure 3A**, initially FIXa and FX bind PSer. This binding reaction is calcium-dependent for all GLA domain-containing proteins. FIXa possesses enzymatic activity to convert X to Xa, whereas in the presence of its cofactor (FVIIIa), the catalytic activity raises 100,000-fold.

As for the FV—there are two different probabilities of binding to the procoagulant surface, as demonstrated in **Figure 3B**. In the first one, prothrombinase forms by the surface Xa reacting with factor Va in solution. Whereas, in the second case, both FVa and FXa form binary complexes with PSer first, and then lipid-protein rearrangement leads to prothrombinase formation [50]. The addition of exogenous FVa to procoagulant platelets increases the velocity of prothrombinase reaction [4], whereas in the absence of exogenous FVa, procoagulant platelets are still prothrombinase active [54]. These findings indicate that both pathways are physiologically important.

3.2 Limiting thrombus growth

During hemostasis, there is a time point when a thrombus needs to stop growing in order not to occlude the lumen of the vessel, which may compromise the blood supply to an end organ. Stoppage of the thrombus growth is probably the most intriguing part of thrombogenesis, although being the most understudied at the same time. When coagulation is initiated by subendothelial tissue factor, it leads to the formation of small amounts of thrombin. Thrombin then, via a positive feedback mechanism, by activating FXI and FVIII triggers the contact activation pathway and amplifies its production. This self-accelerating process is essential within the hemostatic plaque as it is the only way to overcome the anticoagulant nature of plasma, which is due to the presence of antithrombin III (ATIII). Besides the presence of ATIII in the active form, other mechanisms limit coagulation beyond its border zone. Another one of high importance is the presence of thrombomodulin, an integral membrane protein that is expressed ubiquitously on the surface of endothelial cells. Thrombomodulin converts procoagulant thrombin to an anticoagulant enzyme. Not only these mechanisms limit coagulation to the injured site but also degrade activated factors that happened to escape the hemostatic plaque.

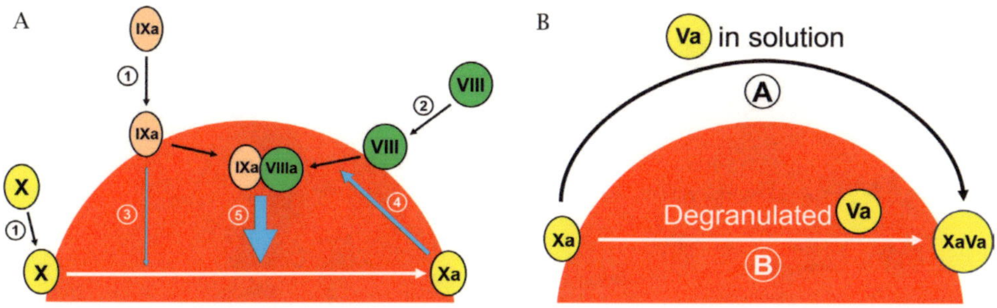

Figure 3.
Factor X activation (A) and prothrombinase assembly (B) on the "cap" of procoagulant platelet. (A). Initially FIXa and FX are bound to PSer on the surface of procoagulant platelet in calcium-dependent manner (①), whereas co-factor VIII binding is not specific to PSer and does not require calcium (②). FIXa possesses enzymatic activity to convert X to Xa (③). Although reaction ③ is very slow in the absence of factor VIIIa, it is efficient to generate small amounts of FXa. Initial FXa then activates VIII to VIIIa (④), this leads to intrinsic tenase complex assembly, which in turn amplifies Xa formation (⑤). (B). There are two possibilities for the assembly of prothrombinase on the surface of procoagulant platelet. Pathway Ⓐ forms prothrombinase by the surface Xa reacting with factor Va in solution. Pathway Ⓑ depicts the possibility for prothrombinase complex formation by already bound Xa and Va. In this case, both FVa and FXa form binary complexes with PSer first and then lipid-protein rearrangement leads to prothrombinase formation.

These processes outline the general principles of limiting coagulation to its border zone. However, not only coagulation but also processes of platelet adhesion, aggregation, and activation have to stop in order at a certain timepoint. How and when thrombus stops growing concerning cellular component had been a mystery for a long time.

In 2007, Maroney et al. demonstrated that procoagulant platelets express active tissue factor pathway inhibitor (TFPI) on their surface [55]. But why platelets expressing procoagulant PSer would also need a strong anticoagulant (TFPI) on

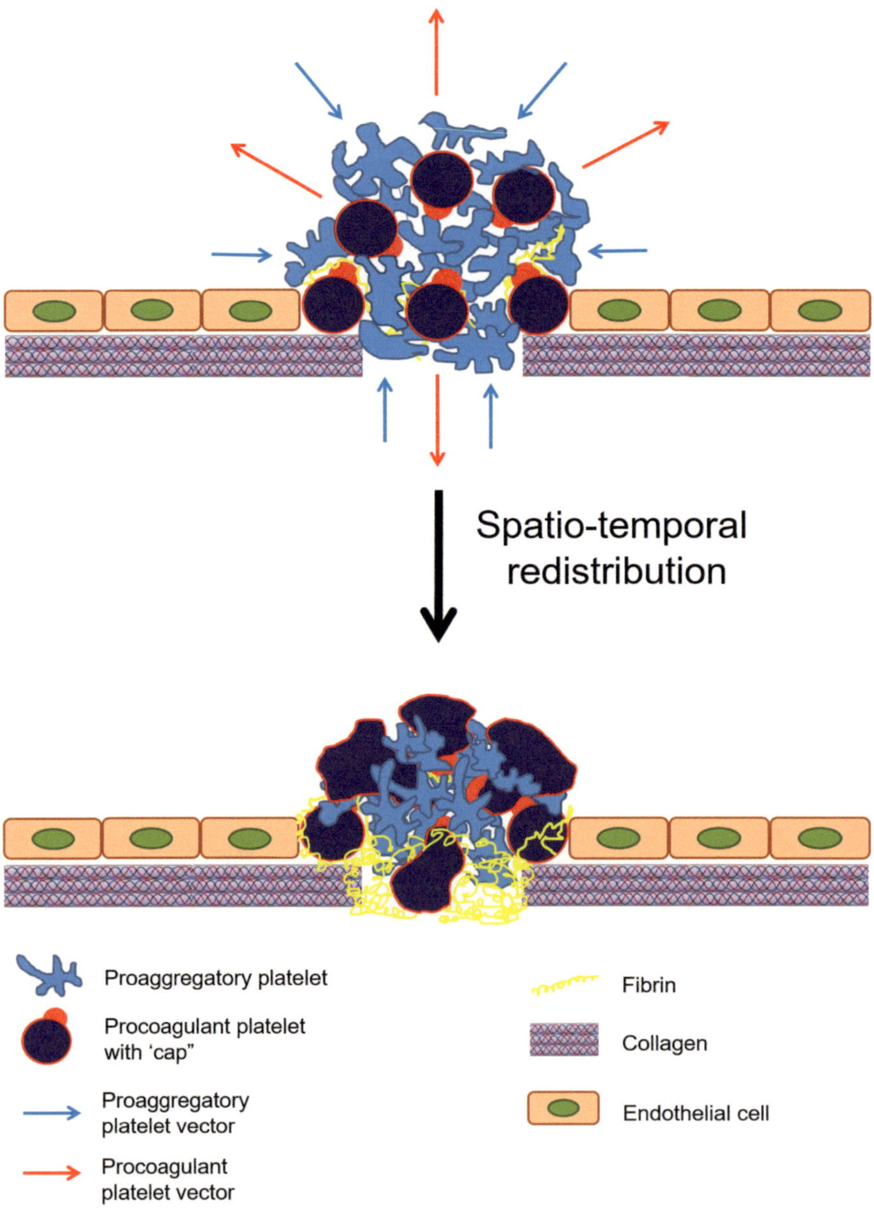

Figure 4.
Schematic representation of platelet translocation within the evolving hemostatic plug. After vascular injury, platelets adhere and aggregate at the wound site. A subpopulation of platelets within the hemostatic plug transitions to a procoagulant state. Procoagulant and pro-aggregatory platelets have opposing vectors of translocation. Clot retraction, driven by platelets with a pro-aggregatory phenotype, squeezes procoagulant platelets to the periphery. Adapted from [56] with minor modification.

their surface remained a question for about a decade until in 2019 a breakthrough work by Nechipurenko et al. characterized a phenomenon, together with its mechanisms, of procoagulant platelets translocating to the thrombus periphery [57]. They demonstrated that during clot retraction procoagulant platelets are squeezed out to the periphery of the thrombus, as shown in **Figure 4**. Another important study in this context is the 2008 work by Jobe et al., which demonstrated that procoagulant platelets do not possess contractile function [18]. This explains why pro-aggregatory and procoagulant platelets have different translocation vectors within the hemostatic plaque (**Figure 4**). Being bound to the thrombus by its "cap," procoagulant platelets do not get detached but rather are squeezed out during pro-aggregatory platelets contraction. At the luminal surface of the hemostatic plaque, procoagulant platelets limit its further growth not only by expressing low adhesive and aggregatory surfaces [58–60], but also with TFPI terminating any extrinsic tenase and prothrombinase activity.

The fact of procoagulant platelet being non-adhesive, however, gives rise to a legitimate question. If procoagulant platelets are not capable of adhesion and aggregation, how do they get attracted to a thrombus? It turns out to be, as described in recent studies, that procoagulant platelets do not form *de novo* from a resting state, but rather temporally transitioning from pro-aggregatory phenotype by the onset of the MPT and regulated necrosis within a hemostatic plaque [61, 62].

Many are aware of an ancient Roman god of duality — Janus, who had two faces as he was able to look to the future and the past. Procoagulant platelets can also demonstrate a functional duality by supporting coagulation on one side and limiting thrombus growth on the other, making them be "Janus" of hemostasis.

Conflict of interest

The author has no conflict of interest to disclose.

Abbreviations

APC	activated protein C
ATIII	antithrombin III
Cyt C	cytochrome c
DTS	dense tubular system
FcRγ	FC receptor gamma
FV	factor V
FVa	activated factor V
FVIII	factor VIII
FVIIIa	activated factor VIII
FIX	factor IX
FIXa	activated factor IX
FX	factor X
FXa	activated factor X
GPIb-IX	glycoprotein Ib and IX complex
GPVI	glycoprotein VI
MCU	mitochondrial calcium uniporter
MPT	mitochondrial permeability transition
MPTP	mitochondrial permeability transition pore
OCS	open canalicular system
PAR	protease-activated receptor

PC	protein C
PLA_2	phospholipase A_2
PS	protein S
PSer	phosphatidylserine
RBC	red blood cell
SOCE	store-operated calcium entry
TF	tissue factor
TFPI	tissue factor pathway inhibitor
TRPC3	transient receptor potential channel 3
WBC	white blood cell

Author details

Andaleb Kholmukhamedov
Versiti Blood Research Institute, Milwaukee, Wisconsin, USA

*Address all correspondence to: akholmukhame@ilww.com

References

[1] Howell W. Nature and action of thromboplastic (zymoplastic) substances of tissues. American Journal of Physiology. 1912;**31**:1-21

[2] Slotta K. Thromboplastin I. Phospholipid moiety of Thromboplastin. Proceedings of the Society for Experimental Biology and Medicine. January 1960;**103**:53-56

[3] Bevers EM, Comfurius P, van Rijn JL, Hemker HC, Zwaal RF. Generation of prothrombin-converting activity and the exposure of phosphatidylserine at the outer surface of platelets. European Journal of Biochemistry. 1982;**122**(2):429-436

[4] Bevers EM, Comfurius P, Hemker HC, Zwaal RFA. On the clot-promoting activity of human platelets in a one-stage prothrombinase assay. Pathophysiology of Haemostasis and Thrombosis. 1982;**12**(3):268-274

[5] Rosing J, van Rijn JL, Bevers EM, van Dieijen G, Comfurius P, Zwaal RF. The role of activated human platelets in prothrombin and factor X activation. Blood. 1985;**65**(2):319-332

[6] Dachary-Prigent J, Freyssinet JM, Pasquet JM, Carron JC, Nurden AT. Annexin V as a probe of aminophospholipid exposure and platelet membrane vesiculation: A flow cytometry study showing a role for free sulfhydryl groups. Blood. 1993;**81**(10):2554-2565

[7] Heemskerk JW, Vuist WM, Feijge MA, Reutelingsperger CP, Lindhout T. Collagen but not fibrinogen surfaces induce bleb formation, exposure of phosphatidylserine, and procoagulant activity of adherent platelets: Evidence for regulation by protein tyrosine kinase-dependent Ca2+ responses. Blood. 1997;**90**(7):2615-2625

[8] Trotter PJ, Orchard MA, Walker JH. Thrombin stimulates the intracellular relocation of annexin V in human platelets. Biochimica et Biophysica Acta. 1994;**1222**(2):135-140

[9] Trotter PJ, Orchard MA, Walker JH. Ca2+ concentration during binding determines the manner in which annexin V binds to membranes. The Biochemical Journal. 1995;**308** (Pt 2):591-598

[10] Tzima E, Poujol C, Nurden P, Nurden AT, Orchard MA, Walker JH. Annexin V relocates to the periphery of activated platelets following thrombin activation: An ultrastructural immunohistochemical approach. Cell Biology International. 1999;**23**(9):629-635

[11] Smeets EF, Heemskerk JW, Comfurius P, Bevers EM, Zwaal RF. Thapsigargin amplifies the platelet procoagulant response caused by thrombin. Thrombosis and Haemostasis. 1993;**70**(6):1024-1029

[12] Zwaal RF, Bevers EM. Platelet phospholipid asymmetry and its significance in hemostasis. Sub-Cellular Biochemistry. 1983;**9**:299-334

[13] Alberio L, Safa O, Clemetson KJ, Esmon CT, Dale GL. Surface expression and functional characterization of alpha-granule factor V in human platelets: Effects of ionophore A23187, thrombin, collagen, and convulxin. Blood. 2000;**95**(5):1694-1702

[14] Kulkarni S, Jackson SP. Platelet factor XIII and calpain negatively regulate integrin alphaIIbbeta3 adhesive function and thrombus growth. The Journal of Biological Chemistry. 2004;**279**(29):30697-30706

[15] Dale GL, Friese P, Batar P, Hamilton SF, Reed GL, Jackson KW,

et al. Stimulated platelets use serotonin to enhance their retention of procoagulant proteins on the cell surface. Nature. 2002;**415**(6868):175-179

[16] Remenyi G, Szasz R, Friese P, Dale GL. Role of mitochondrial permeability transition pore in coated-platelet formation. Arteriosclerosis, Thrombosis, and Vascular Biology. 2005;**25**(2):467-471

[17] Panteleev MA, Ananyeva NM, Greco NJ, Ataullakhanov FI, Saenko EL. Two subpopulations of thrombin-activated platelets differ in their binding of the components of the intrinsic factor X-activating complex. Journal of Thrombosis and Haemostasis. 2005;**3**(11):2545-2553

[18] Jobe SM, Wilson KM, Leo L, Raimondi A, Molkentin JD, Lentz SR, et al. Critical role for the mitochondrial permeability transition pore and cyclophilin D in platelet activation and thrombosis. Blood. 2008;**111**(3):1257-1265

[19] Agbani EO, van den Bosch MTJ, Brown E, Williams CM, Mattheij NJA, Cosemans JMEM, et al. Coordinated membrane ballooning and procoagulant spreading in human platelets. Circulation. 2015;**132**(15):1414-1424

[20] Mazepa M, Hoffman M, Monroe D. Superactivated platelets: Thrombus regulators, thrombin generators, and potential clinical targets. Arteriosclerosis, Thrombosis, and Vascular Biology. 2013;**33**(8):1747-1752

[21] Jobe SM, Leo L, Eastvold JS, Dickneite G, Ratliff TL, Lentz SR, et al. Role of FcRγ and factor XIIIA in coated platelet formation. Blood. 2005;**106**(13):4146-4151

[22] Choo H-J, Kholmukhamedov A, Zhou C, Jobe S. Inner mitochondrial membrane disruption links apoptotic and agonist-initiated phosphatidylserine externalization in platelets. Arteriosclerosis, Thrombosis, and Vascular Biology. 2017;**37**(8):1503-1512

[23] Kim J-S, He L, Lemasters JJ. Mitochondrial permeability transition: A common pathway to necrosis and apoptosis. Biochemical and Biophysical Research Communications. 2003;**304**(3):463-470

[24] Hunter DR, Haworth RA, Southard JH. Relationship between configuration, function, and permeability in calcium-treated mitochondria. The Journal of Biological Chemistry. 1976;**251**(16):5069-5077

[25] Bernardi P, Forte M. The mitochondrial permeability transition pore. Novartis Foundation Symposium. 2007;**287**:157-164. Discussion 164-169

[26] Hathaway DR, Adelstein RS. Human platelet myosin light chain kinase requires the calcium-binding protein calmodulin for activity. Proceedings of the National Academy of Sciences of the United States of America. 1979;**76**(4):1653-1657

[27] Shattil SJ, Brass LF. Induction of the fibrinogen receptor on human platelets by intracellular mediators. The Journal of Biological Chemistry. 1987;**262**(3):992-1000

[28] Kholmukhamedov A, Janecke R, Choo H-J, Jobe SM. The mitochondrial calcium uniporter regulates procoagulant platelet formation. Journal of Thrombosis and Haemostasis. 2018;**16**(11):2315-2321

[29] Podoplelova NA, Sveshnikova AN, Kotova YN, Eckly A, Receveur N, Nechipurenko DY, et al. Coagulation factors bound to procoagulant platelets concentrate in cap structures to promote clotting. Blood. 2016;**128**(13):1745-1755

[30] Kholmukhamedov A, Jobe S. Necrotic but not apoptotic platelets are functionally procoagulant. Blood. 2018;**132**(Suppl 1):2420

[31] McArthur K, Chappaz S, Kile BT. Apoptosis in megakaryocytes and platelets: The life and death of a lineage. Blood. 2018;**131**(6):605-610

[32] Pleines I, Lebois M, Gangatirkar P, Au AE, Lane RM, Henley KJ, et al. Intrinsic apoptosis circumvents the functional decline of circulating platelets but does not cause the storage lesion. Blood. 2018;**132**(2):197-209

[33] Mason KD, Carpinelli MR, Fletcher JI, Collinge JE, Hilton AA, Ellis S, et al. Programmed anuclear cell death delimits platelet life span. Cell. 2007;**128**(6):1173-1186

[34] Josefsson EC, James C, Henley KJ, Debrincat MA, Rogers KL, Dowling MR, et al. Megakaryocytes possess a functional intrinsic apoptosis pathway that must be restrained to survive and produce platelets. The Journal of Experimental Medicine. 2011;**208**(10):2017-2031

[35] Prodan CI, Vincent AS, Dale GL. Coated-platelet levels increase with number of injuries in patients with mild traumatic brain injury. Journal of Neurotrauma. 2016;**33**(9):818-824

[36] Prodan CI, Stoner JA, Dale GL. Acute hemorrhagic complications are associated with lower coated-platelet levels in non-lacunar brain infarction. Journal of Thrombosis and Haemostasis. 2015;**13**(12):2233-2239

[37] Prodan CI, Stoner JA, Dale GL. Lower coated-platelet levels are associated with increased mortality after spontaneous intracerebral hemorrhage. Stroke. 2015;**46**(7):1819-1825

[38] Prodan CI, Vincent AS, Dale GL. Coated-platelet levels are persistently elevated in patients with mild traumatic brain injury. The Journal of Head Trauma Rehabilitation. 2014;**29**(6):522-526

[39] Kirkpatrick AC, Vincent AS, Dale GL, Prodan CI. Coated-platelets predict stroke at 30 days following TIA. Neurology. 2017;**89**(2):125-128

[40] Kirkpatrick AC, Stoner JA, Dale GL, Rabadi M, Prodan CI. Higher coated-platelet levels in acute stroke are associated with lower cognitive scores at three months post infarction. Journal of Stroke and Cerebrovascular Diseases. 2019;**28**(9):2398-2406

[41] Kirkpatrick AC, Tafur AJ, Vincent AS, Dale GL, Prodan CI. Coated-platelets improve prediction of stroke and transient ischemic attack in asymptomatic internal carotid artery stenosis. Stroke. 2014;**45**(10):2995-3001

[42] Ray B, Pandav VM, Mathews EA, Thompson DM, Ford L, Yearout LK, et al. Coated-platelet trends predict short-term clinical outcome after subarachnoid hemorrhage. Translational Stroke Research. 2018;**9**(5):459-470

[43] Sakurai Y, Hardy ET, Ahn B, Tran R, Fay ME, Ciciliano JC, et al. A microengineered vascularized bleeding model that integrates the principal components of hemostasis. Nature Communications. 2018;**9**(1):509

[44] Kholmukhamedov A, Jobe S. Mitochondrial uncoupling prevents procoagulant platelet formation resulting in decreased thrombus stability. Blood. 2018;**132**(Suppl 1):2421

[45] Panteleev MA, Saenko EL, Ananyeva NM, Ataullakhanov FI. Kinetics of factor X activation by the membrane-bound complex of factor IXa and factor VIIIa. The Biochemical Journal. 2004;**381**(Pt 3):779-794

[46] Ataullakhanov FI, Panteleev MA. Mathematical modeling and computer simulation in blood coagulation. Pathophysiology of Haemostasis and Thrombosis. 2005;**34**(2-3):60-70

[47] Panteleev MA, Balandina AN, Lipets EN, Ovanesov MV, Ataullakhanov FI. Task-oriented modular decomposition of biological networks: Trigger mechanism in blood coagulation. Biophysical Journal. 2010;**98**(9):1751-1761

[48] Ovanesov MV, Ananyeva NM, Panteleev MA, Ataullakhanov FI, Saenko EL. Initiation and propagation of coagulation from tissue factor-bearing cell monolayers to plasma: Initiator cells do not regulate spatial growth rate. Journal of Thrombosis and Haemostasis. 2005;**3**(2):321-331

[49] Sinauridze EI, Kireev DA, Popenko NY, Pichugin AV, Panteleev MA, Krymskaya OV, et al. Platelet microparticle membranes have 50- to 100-fold higher specific procoagulant activity than activated platelets. Thrombosis and Haemostasis. 2007;**97**(3):425-434

[50] Krishnaswamy S, Nesheim ME, Pryzdial EL, Mann KG. Assembly of prothrombinase complex. Methods in Enzymology. 1993;**222**:260-280

[51] Mann KG, Krishnaswamy S, Lawson JH. Surface-dependent hemostasis. Seminars in Hematology. 1992;**29**(3):213-226

[52] Gilbert GE, Novakovic VA, Shi J, Rasmussen J, Pipe SW. Platelet binding sites for factor VIII in relation to fibrin and phosphatidylserine. Blood. 2015;**126**(10):1237-1244

[53] Gilbert GE, Arena AA. Activation of the factor VIIIa-factor IXa enzyme complex of blood coagulation by membranes containing phosphatidyl-L-serine. The Journal of Biological Chemistry. 1996;**271**(19):11120-11125

[54] Nesheim ME, Nichols WL, Cole TL, Houston JG, Schenk RB, Mann KG, et al. Isolation and study of an acquired inhibitor of human coagulation factor V. The Journal of Clinical Investigation. 1986;**77**(2):405-415

[55] Maroney SA, Haberichter SL, Friese P, Collins ML, Ferrel JP, Dale GL, et al. Active tissue factor pathway inhibitor is expressed on the surface of coated platelets. Blood. 2007;**109**(5):1931-1937

[56] Kholmukhamedov A, Jobe S. Procoagulant platelets get squeezed to define the boundaries of the hemostatic plug. Arteriosclerosis, Thrombosis, and Vascular Biology. 2019;**39**(1):5-6

[57] Nechipurenko DY, Receveur N, Yakimenko AO, Shepelyuk TO, Yakusheva AA, Kerimov RR, et al. Clot contraction drives the translocation of procoagulant platelets to thrombus surface. Arteriosclerosis, Thrombosis, and Vascular Biology. 2019;**39**(1):37-47

[58] Mattheij NJA, Gilio K, van Kruchten R, Jobe SM, Wieschhaus AJ, Chishti AH, et al. Dual mechanism of integrin αIIbβ3 closure in procoagulant platelets. The Journal of Biological Chemistry. 2013;**288**(19):13325-13336

[59] Liu F, Gamez G, Myers DR, Clemmons W, Lam WA, Jobe SM. Mitochondrially mediated integrin αIIbβ3 protein inactivation limits thrombus growth. The Journal of Biological Chemistry. 2013;**288**(42):30672-30681

[60] Yakimenko AO, Verholomova FY, Kotova YN, Ataullakhanov FI, Panteleev MA. Identification of different proaggregatory abilities of activated platelet subpopulations. Biophysical Journal. 2012;**102**(10):2261-2269

[61] Alberio L, Ravanat C, Hechler B, Mangin PH, Lanza F, Gachet C. Delayed-onset of

procoagulant signalling revealed by kinetic analysis of COAT platelet formation. Thrombosis and Haemostasis. 2017;**117**(6):1101-1114

[62] Kholmukhamedov A, Jobe S. Mitochondria and platelet cell death. Thrombosis and Haemostasis. 2017;**117**(11):2207-2208

Chapter 4

MicroRNAs in Platelets: Should I Stay or Should I Go?

Sonia Águila, Ernesto Cuenca-Zamora, Constantino Martínez and Raúl Teruel-Montoya

Abstract

In this chapter, we discuss different topics always using the microRNA as the guiding thread of the review. MicroRNAs, member of small noncoding RNAs family, are an important element involved in gene expression. We cover different issues such as their importance in the differentiation and maturation of megakaryocytes (megakaryopoiesis), as well as the role in platelets formation (thrombopoiesis) focusing on the described relationship between miRNA and critical myeloid lineage transcription factors such as RUNX1, chemokines receptors as CRCX4, or central hormones in platelet homeostasis like TPO, as well as its receptor (MPL) and the TPO signal transduction pathway, that is JAK/STAT. In addition to platelet biogenesis, we review the microRNA participation in platelets physiology and function. This review also introduces the use of miRNAs as biomarkers of platelet function since the detection of pathogenic situations or response to therapy using these noncoding RNAs is getting increasing interest in disease management. Finally, this chapter describes the participation of platelets in cellular interplay, since extracellular vesicles have been demonstrated to have the ability to deliver microRNAs to others cells, modulating their function through intercellular communication, redefining the extracellular vesicles from the so-called "platelet dust" to become mediators of intercellular communication.

Keywords: platelets, megakaryocytes, microRNAs, intercellular communication, microvesicles

1. Introduction

The discovery of noncoding RNAs, apart from raising new questions about the central dogma of molecular biology, has greatly contributed to the basic knowledge of numerous diseases, and to their therapeutic utility [1]. Nowadays, nobody doubts that the so-called "junk genome," a large DNA proportion of the genome [2], undoubtedly participates in the flow of genomic information [3] that gives rise to cellular functions and that frequent diseases are associated with malfunction of noncoding RNAs [4].

Although essential processes are common to all kind of cells, in physiological condition, each tissue or cell has unique features or responses that define their phenotype. These functions will be controlled by the gene expression patterns of that tissue and therefore by the regulatory mechanisms of such expression.

Thus microRNAs (miRNAs), which are a family of small noncoding RNAs, are an important element in gene-expression-regulation [5]. This function of miRNAs, which is exerted through repression of gene expression, is related to the specific miRNA expression, in a spacial and timely manner; in other words, it is tissue and/or developmental stage-specific [6].

This becomes more important in a cell that lacks a nucleus, such as platelets, and therefore lacks the ability to regulate gene expression through canonical mechanisms such as transcription factors (TFs) or epigenetic mechanisms of genome modification, for example, cytosine methylation or histone acetylation. In this scenario, miRNAs may acquire more prominence in the processes of gene regulation at a post-transcriptional level [7], not only during platelet production (thrombopoiesis), but also providing to the platelets with certain capacity to modulate their phenotype and consequently contribute to their ability to respond to external stimuli.

Historically, the study of hematopoiesis has been focused in general on the investigation of cytokine receptors and TFs, which ultimately govern the different membrane marker characteristics of each hematopoietic lineage, in a continuous change in gene expression patterns. The TFs expressed by megakaryocyte (MK) progenitors that enable their engagement with the lineage are increasingly understood. During MK differentiation, the progression to specific hematopoietic pathways is influenced by changes in the levels of these factors. These changes are dynamic, and in this scenario, it is clear that the repressive action exerted by miRNAs on their targets, for example, TFs, may be one more element to pay attention to in hematopoiesis [8, 9].

Finally, as mentioned above, there is a certain specificity of miRNA expression at the tissue-cell level, but in recent years cellular communication mechanisms have also been described in which a transfer of content from cell to cell [10], through extracellular vesicles (EVs), for example, microvesicles or exosomes, can occur. Within the content of EVs, we can find genetic material such as miRNAs [11], thus making possible the regulation of gene expression at a distance. If we add to this the fact that platelets provide a large proportion of the total cellular miRNAs found in blood [12], in addition to the high correlation observed in miRNA profiles expression between plasma and platelets [13] and that most of the EVs that circulate in bloodstream come from platelets [14], their study from a pathophysiological point of view, as well as their possible use as a therapeutic tool, is very relevant.

In summary, in this chapter we intend to discuss different topics, always using the miRNA as the guiding thread of the review, which would cover issues such as their importance in the platelet biogenesis, as well as their participation in their (platelets') physiology and function. This review will also introduce issues such as the use of miRNAs as biomarkers of platelet function, and the detection of pathogenic situations. Finally, this chapter will describe the participation of platelets in cellular interplay, since EVs have been demonstrated to have the ability to deliver their cargo to other cells, modulating their function through intercellular communication.

2. Biogenesis and function of miRNA

2.1 Biogenesis

The synthesis of miRNAs begins in the nucleus, through their transcription by RNA polymerase II, giving rise to a product called pri-miRNA, which has a size range from several hundred nucleotides (nt), to several kilobases; this may depend on the miRNA gene location, that is, polycistronic, intronic, or introgenic. Still in the nucleus, pri-miRNA is processed into pre-miRNA by the Microprocessor catalytic complex, composed by Drosha and DGCR8. The first component is a nuclease with two RNAse III domains, which are responsible for cutting the pri-miRNAs. DGCR8, on the other hand, contains RNA-binding domains that stabilize the pri-miRNA for being processed by Drosha [15]. The pre-miRNA resulting structure is 3'overhang hairpin-like shape and it has a size of approximately 70 nt. Pre-miRNA will leave the nucleus, mainly by a RanGTP/export 5-dependent mechanism and once in the cytoplasm will be reprocessed, by Dicer, resulting in a new ~22 nt size double-stranded form [16]. Of the two strands that form the mature-duplex pre-miRNA, one is the so-called "guide" strand and will be loaded into Argonaute (AGO), while the other strand may be degraded or even become functional too. Despite the fact that general aspects of the miRNA maturation pathway have been known for a long time, at present and due to structural studies, the functions of the macromolecular complexes involved are increasingly known in detail, giving a deeper understanding of their involvement in miRNA cytoplasmic processing [16, 17].

2.2 Function

The complete functional unit consists of the mature miRNA and the AGO protein, which recognizes by base pairing its target mRNA, mainly in the 3′ untranslated (3'UTR) region, and this union will induce both, an inhibition of mRNA translation in its first stages and a RNA decaying. The repressive action on translation is exerted by the initiation factors 4 A-I (eIF4A-I) and eIF4A-II. On the other hand, the mRNA decay is favored by the 5'mRNA deadenylation, followed by its decapping [18]. Finally, this leads to an indelible genetic silencing, which is often complex to study due to the intricate network and interactions that occur between different miRNAs and mRNAs, since a miRNA can have multiple target mRNAs, and a single mRNA can be cooperatively regulated by multiple miRNAs (**Figure 1**).

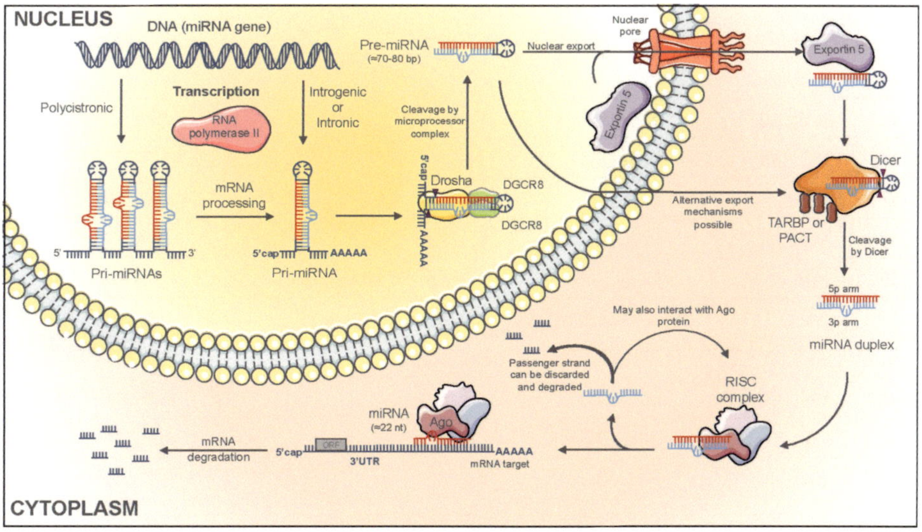

Figure 1.
Outline of the processes involved both in miRNA biogenesis and in its function.

3. Platelets miRNA repository

Since Landry et al. showed in 2009 that platelets have abundant and diverse miRNAs [19], numerous studies have analyzed the content of miRNA in platelets. We review the miRNAs detected in platelets, focusing especially on those detected with greater expression. First, we have to keep in mind that there are different technologies, therefore platforms, in which a high throughput miRNA analysis can be performed, such as microarray, nanostring, or RNA-seq [20]. We will not go into details about the technologies used, but we will rather give importance to the experimental conditions used, in terms of the platelets purity. Since the total RNA content in platelets, in absolute terms, is very low, compared to other blood cells [12], that is leucocytes, a possible contamination can alter or bias the results. Thus, we here focus on studies that performed a leukocyte depletion step by filtration or immunoselection. We summarize in a table (**Table 1**), the expression of the 40 most expressed miRNAs from selected studies, in order to outline the miRNAs that coincide between them. To note, all the provided information comes from human

	Landry [19]	**Nagalla [21]**	**Edelstein [22] and Simon [23]**	**Teruel-Montoya [24]**
1	miR-142-5p	miR-223	miR-223	miR-126
2	miR-142-3p	miR-26b	miR-26a	miR-223
3	miR-223	miR-26a	miR-126	miR-142-3p
4	let-7a	miR-23a	miR-142-3p	miR-26a
5	miR-185	miR-126	miR-16	let-7 g
6	let-7c	miR-21	miR-92a	miR-16
7	let-7i	let-7f	miR-21	miR-15b
8	let-7b	miR-22	miR-103a	miR-92a
9	miR-126	miR-24	miR-20a/b	let-7a
10	miR-103a	miR-720	let-7a	let-7d
11	miR-320	miR-16	miR-24	miR-21
12	miR-30c	miR-23b	let-7 g	miR-199a/b
13	miR-130a	miR-142-3p	miR-199a/b	miR-451a
14	miR-26a	miR-142-5p	miR-15a	miR-15a
15	miR-191	miR-191	let-7d	let-7f
16	miR-30b	miR-451a	miR-30b	miR-23a
17	miR-146a	miR-30b	let-7f	miR-103a
18	miR-23b	miR-15b	let-7i	let-7i
19	miR-21	let-7a	miR-15b	miR-20a/b
20	miR-23a	miR-1826	miR-23a	miR-24
21	miR-19a	miR-17	miR-221	miR-191
22	miR-106a	miR-103a	miR-146a	miR-25
23	miR-15b	miR-106a	miR-451a	miR-19b
24	miR-20a	miR-30c	miR-19b	miR-30b
25	miR-107	miR-130a	miR-142-5p	miR-221
26	miR-451a	let-7 g	miR-148b	miR-148b
27	miR-222	miR-30a	miR-191	miR-146a

	Landry [19]	Nagalla [21]	Edelstein [22] and Simon [23]	Teruel-Montoya [24]
28	miR-151-5p	miR-146b	miR-25	miR-27b
29	miR-24	miR-34b	miR-106b	miR-151a
30	miR-625	miR-106b	miR-423-5p	miR-181a
31	miR-671	miR-15a	miR-27b	miR-720
32	miR-29a	miR-374a	miR-340	miR-29a
33	miR-22	miR-20a	miR-181a	miR-142-5p
34	miR-16	miR-185	miR-151-5p	miR-423
35	miR-199a/b	miR-146a	miR-720	miR-125a
36	miR-106b	let-7i	miR-199a-5p	miR-374a
37	miR-146b	let-7c	miR-101	miR-106b
38	miR-151-3p	miR-19a	miR-374b	miR-29c
39	miR-98	let-7d	miR-29a	miR-335
40	miR-148b	miR-27a	miR-148a	miR-148a

Underlined those miRNAs that appear in the top 40 in all the studies.

Table 1.
Comparison of the 40 most expressed miRNA in platelets ranked by expression.Rank.

healthy subjects. While it is true that there may be differences in age, gender, or race [22, 23], we observe that there are 18 miRNA coincidences, that is, miR-223, miR-126, miR-26a, miR-142-3p, miR-16, miR-21, miR-103a, let-7a, miR-24, miR-30b, let-7i, miR-15b, miR-23a, miR-146a, miR-451a, miR-142-5p, miR-191, and miR-106b. With a lower degree of coincidence, there are five miRNA coincidences in four out of the five publications: let-7f, let-7d, miR-15a, let-7 g, and miR-720. In addition, there are 11 miRNA matches in at least 3 of the 5 datasets, that is, miR-103, miR-19b, miR-27b, miR-92a, miR-181a, miR-148a, miR-221, miR-25, miR-20a/b, and miR-151-5p.

The knowledge of the platelets-miRNA-cargo, as it is explained throughout the chapter, is important for understanding not only platelet biogenesis and physiology itself, but also MK function. Thus, platelets represent a faithful reflection of the physiological state of its progenitor with the advantage that platelets are much more accessible than the MKs. A better understanding of the physiological state of the platelet will undoubtedly help us to detect oscillations that may be pathological, and it is here that miRNAs can help us in diagnosis, prognosis, as well as to follow-up the disease progression. In addition to their great accessibility, circulating and platelet miRNAs are relatively easy to measure [25] and their physical-chemical properties make them good candidates for being used as biomarkers [26]. Furthermore, platelets are an important source of circulating miRNAs that can remotely regulate gene expression in other cells, such as macrophages or endothelial cells.

4. Role of miRNA in megakaryopoiesis and thrombopoiesis

Megakaryopoiesis and subsequent thrombopoiesis occur through a series of complex biological processes. First, MK precursors developed from hematopoietic stem cells (HSCs) initially proliferate, and then differentiate into mature polyploid MKs, which eventually release platelets. Thus, we cannot ignore great peculiarities of platelets, in particular the fact that they are anucleate cells derived from MKs and that their content is the reflection of their parental cells, including the miRNA cargo. Accordingly, we consider it essential to describe the role of miRNAs in platelet and the megakaryopoiesis itself.

4.1 miRNAs in megakaryopoiesis

Megakaryopoiesis takes place primarily in the bone marrow. To understand this process in a simple way, it can be separated into two phases: the proliferative phase, in which the expansion of the MK precursors takes place, and the maturation phase, in which the two main events of this lineage occur. The first is at the nuclear level, the polyploidization by endomitosis, and the second is the cytoplasmic maturation [27].

In the last decade, many miRNAs have been described that can regulate or be regulated by TFs implicated in megakaryopoiesis (**Table 2**). For practical reasons, we only focus on some of the regulatory miRNAs. The miR-144/451 cluster, the miRNA most expressed in erythrocytes [12], can be repressed by RUNX1 during megakaryopoiesis. In myeloid differentiation, RUNX1 takes on a major role, since it represses erythroid lineage-specific genes and at the same time activates the transcription of specific genes of the MK lineage [28], and it is known to regulate MK polyploidization [29] and cytoskeleton rearrangement in the process of MK maturation formation [30]. On the other hand, it has been shown that miR-144/451 locus is activated by GATA1 [31–33], which primarily promotes erythropoiesis, but which has also been described as important for megakaryopoiesis. Another well-studied TF-miRNA interaction is mirR-27a and Runx1. miR-27a can repress Runx1 expression in mice. During megakaryopoiesis, Runx1 and miR-27a are engaged into a feedback positive loop regulation of miR-27a expression by Runx1. Ben-ami et al. also observed in K562 a human immortalized myelogenous leukemia cell line, and upon megakaryocytic differentiation by 12-o-tetradecanoylphorbol-13-acetate (TPA), that RUNX1 binds to a putative miR-27a regulatory region and upregulates its expression [34]. RUNX1-miRNA relationship goes in the direction of miRNA regulation by RUNXI, but RUNXI can also be regulated by miRNA; in this sense, miR-9 can regulate the expression of RUNX1. The increase of this miRNA in human MEG-01 and DAMI cell lines, both megakaryoblast phenotype cell lines, has shown a decrease in RUNX1 at both mRNA and protein levels [35]. Furthermore, an inverse level of expression in RUNX1 and miR-9 has been observed in MKs derived from umbilical cord blood (MKCB) and peripheral blood (MKPB) [35, 36]. We can also mention two more cases: (i) PLZF, a transcription factor whose expression increases during megakaryopoiesis, can downregulate miR-146a, which represses CXCR4 [37]; therefore, PLZF activates CXCR4 translation and the increase in CRCX4 can induce MK migration through bone marrow. (ii) ETS1 and MEIS1, both, are TFs with well-known functions in hematopoiesis; ETS1 is upregulated in megakaryocytopoiesis, regulating MK-specific gene promoters, such as platelet factor 4, GATA-2 or GPIIb, indicating that ETS1 promotes MK differentiation [38], and MEIS1 is vital for megakaryopoiesis and thrombopoiesis from human pluripotent stem cells [39]. MiR-155 downregulates both ETS1 and MEIS1, in HSCs, whereas miR-155 expression rapidly declines during TPO-induced megakaryocytic differentiation [40], thus favoring the process of megakaryocyte differentiation through ETS1 and MEIS1. As we have seen, there is evidence of the possible regulation of megakaryopoiesis by miRNAs, a connection between TF and miRNAs, and that the expression of the miRNA-coding genes is affected [41]. We do not know if this is due to a causal relationship or a consequence of the maturation process, perhaps

TF	miRNA	Function/effect	Ref.
RUNX1	miR-144/451	miR-144/451 cluster is repressed by RUNX1 during megakaryocytopoiesis	[28]
	miR-9	miR-9 represses RUNX1 during megakaryocytic differentiation	[35]
		miR-9 expression is higher in MKs from cord blood than MKs from peripheral blood, as opposed to RUNX1	[35, 42]
	miR-27a	In megakaryocytic differentiation, miR-27a is stimulated by RUNX1, and it can directly target RUNX1 (negative feedback loop)	[34]
FLI1	miR-145	miR-145 downregulates FLI1 during megakaryocytic differentiation. This increases MKs production compared to erythrocytes	[43]
		FLI1 may target back miR-145 promoter, creating a negative feedback loop	[44]
GATA1	miR-144/451	GATA1 activates miR-144/451 locus	[31–33]
	miR-138	miR-138 expression is increased by GATA1, and miR-138 represses BCR-ABL1 fusion gene	[45, 46]
HOXA1	miR-10a	miR-10a represses HOXA1. miR-10 is downregulated in human CD34+ bone marrow progenitor-derived MKs	[47]
		In cord blood HSCs, miR-10a knockdown increases HOXA1 expression and stimulates megakaryocytic differentiation, even in the absence of TPO	[48]
HOXA11	miR-181a	It is predicted that miR-181a targets HOXA11	[49]
ETS1 MEIS1	miR-155	miR-155 downregulates both ETS1 and MEIS1. In human HSPCs, during TPO-induced megakaryocytic differentiation, miR-155 expression is rapidly declined	[40]
ETV6	miR-181a	It is predicted that miR-181a targets ETV6	[49]
		miR-181a targets ETV6/RUNX1 gene fusion, observed in acute lymphoblastic leukemia	[50]
	miR-320a	miR-320a is downregulated by ETV6/RUNX1 gene fusion, and targets survivin, an antiapoptotic protein	[50, 51]
	miR-494	miR-494 is downregulated by ETV6/RUNX1 gene fusion, and targets survivin, an antiapoptotic protein	[50, 51]
MAFB	miR-130a	miR-130a, which represses MAFB, is downregulated during human megakaryocytic differentiation	[47]
GFI1	miR-22	miR-22 downregulates GFI1, stimulating myeloid and lymphoid differentiation	[52, 53]
		Overexpression of miR-22 increased K562 megakaryocytic differentiation, while miR-22 knockout inhibited this process induced by PMA (and upregulates GFI1)	[53]
		miR-22 knockout mice showed less of both immature and mature MKs in bone marrow and more GFI1 expression	[53]
GFI1B	miR-22	GFI1B, a GFI1 paralog without miR-22 seed sequence, is highly expressed in MK lineage and in K562 cells, and promotes MK and erythrocyte differentiation	[53]
		In this case, miR-22 may regulate GFI1B indirectly through GFI1 because GFI1 and GFI1B could compete for DNA occupancy	[53]

TF	miRNA	Function/effect	Ref.
FOSB	miR-22	FOSB stimulates miR-22 expression in K562 stimulated by PMA, required for megakaryocytopoiesis	[57]
		miR-22 expression could be regulated by opposing activities of FOSB and GFI1	[58]
PLZF (ZBTB16)	miR-146a	ZBTB16 downregulates miR-146, which represses CXCR4.	[37]
		miR-146a and CXCR4 (and its ligand SDF1) regulate HSC homing, MK proliferation, differentiation, and maturation	[59, 60]
EVI1	miR-133	EVI1 overexpression upregulates miR-133, in acute myeloid leukemia cells	[61]
		miR-133 downregulates EVI1, making a negative feedback loop	[62]
	miR-1	EVI1 overexpression upregulates miR-1, which increases cell proliferation, in acute myeloid leukemia cells	[61]
	miR-449a	miR-449a is repressed by EVI1, and this microRNA downregulates NOTCH1 and BCL2. In this way, miR-449a expression decreases cell viability and increases apoptosis of MECOM overexpressing leukemic cells	[63]

ABL1: ABL proto-oncogene 1, non-receptor tyrosine kinase; BCL2: BCL2 apoptosis regulator; BCR: BCR activator of RhoGEF and GTPase; CXCR4: C-X-C motif chemokine receptor 4; ETS1: ETS proto-oncogene 1; ETV6: ETS variant transcription factor 6; EVI1: ecotropic viral integration site 1; FLI1: Fli-1 proto-oncogene, ETS transcription factor; FOSB: FosB proto-oncogene, AP-1 transcription factor subunit; GATA1: GATA binding protein 1; GFI1: growth factor independent 1 transcriptional repressor; GFI1B: growth factor independent 1B transcriptional repressor; HOXA1: Homeobox A1; HOXA11: Homeobox A11; HSCs: hematopoietic stem cells; HSPCs: hematopoietic stem and progenitor cells; MAFB: MAF BZIP transcription factor B; MECOM: MDS1 and EVI1 complex locus; MEIS1: Meis Homeobox 1; MKs: megakaryocytes; NOTCH1: notch receptor 1; PLZF: promyelocytic leukaemia zinc finger; PMA: phorbol myristate acetate; RUNX1: RUNX family transcription factor 1; SDF1: stromal cell-derived factor 1; TPO: thrombopoietin; ZBTB16: zinc finger and BTB domain containing 16.

Table 2.
Transcription factors-miRNA relationship in megakaryocytic differentiation.

both, but what we do know is that it can have a side effect on the expression of many other genes.

Besides the relationships of TFs and miRNAs in megakaryopoiesis, other miRNAs have been described as complex effectors of megakaryopoiesis, for example the miR-146 family, which includes miR-146a and miR-146b. Specifically, miR-146a, a miRNA closely related to inflammatory diseases, seems to play an important role in the development of normal hematopoiesis. Thus, deficient miR-146a mice, stands out for the phenotypic features of abnormal hematopoiesis, highlighted by bone marrow myelofibrosis [54]. Its complex role in megakariopoiesis lies in part in some controversy about the role of this miRNA in this process. It has been described that miR-146a expression is high during induced megakaryocytopoiesis in vivo in murine model and in vitro in human cell culture; however, enforced miR-146a expression has minimal effects on the process [41]. On the other hand, it has been reported that miR-146a expression is downregulated when human cord blood-derived CD34+ cells are induced to differentiate into MKs [37], but miR-146a overexpression impaired megakaryocytopoiesis [41] and knockdown of miR-146a in mouse HSCs resulted in increased MKs in the bone marrow [55, 56]. This disagreement may be due to differences in experimental conditions and the differences between human and murine models that have been used, but this does not detract from the fact that miR-146a may actually have an effect on megakaryopoiesis. The other member of miR-146 family, miR-146b, regulates directly and indirectly, through GATA-1, the expression of the platelet-derived growth factor receptor α (PDGFRA), in phorbol 12-myristate 13-acetate (PMA)-differentiated K562 cells. In addition, the expression of miR-146b increases in CD34+ hematopoietic stem/progenitor cells undergoing megakaryocytic differentiation; at the same time, the expression of PDGFRA decreases [64]. Finally and also in CD34+ derived MK differentiation assays, other researchers suggest that miR-28 plays a negative role in the differentiation of MK precursors. Through miR-28 transduction experiments on human CD34+ cells, with subsequent differentiation to MKs, using thrombopoietin (TPO), they observed a reduction of more than 50% in the number of MKs with proplatelets. This is, in part, explained by a repression of MPL gene expression, which encodes for the TPO receptor [65]. In relation with TPO, the expression of miR-150 has been described to increase with this hormone [66], and assays consisting in overexpression of miR-150, seem to suggest that elevated levels of miR-150 enhanced both in vitro and in vivo megakaryocyte differentiation at the expense of erythroid differentiation [67].

4.2 miRNAs in thrombopoiesis

Thrombopoiesis is the process by which platelets are generated from MKs During this process, microtubules mediate elongation of the MK extensions (proplatelet, **Figure 2**) and granule trafficking from MKs to nascent platelets [68].

As we have already mentioned, the TFs cited above are not only of great importance in megakaryopoiesis, but also in thrombopoiesis. It is very intuitive to think that the miRNAs that can regulate these TFs will also affect thrombopoiesis. Since many of the mutations described in RUNX1, FLI1, GATA1, GFI1B, ETV6, EVI1, and HOXA11 have been associated with variable thrombocytopenia [69]. The same effect can be expected for both cases, in the repression of TF-coding-gene expression, and when a mutation is observed that causes a loss of function in the same TF. Therefore, as far as possible, we will not be redundant with those relationships of miRNAs and TFs already described above (**Table 2**) that have repercussions on thrombopoiesis. In addition miRNAs can exert a quantitative regulation of

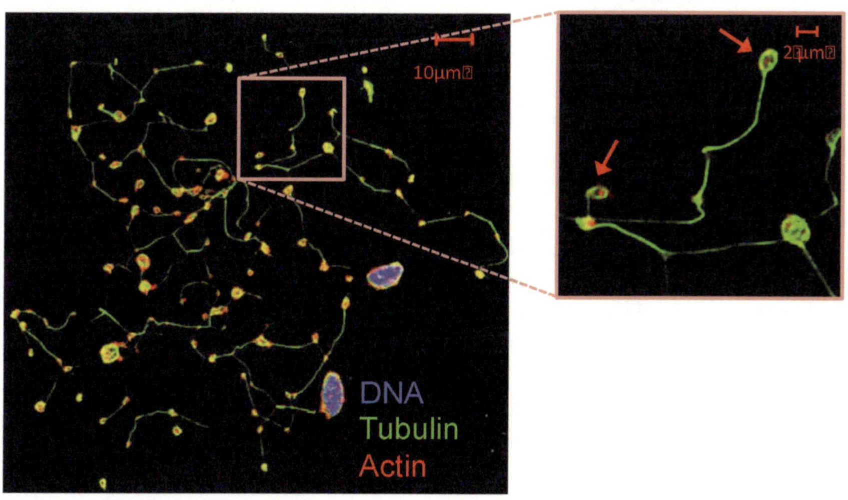

Figure 2.
Megakaryocytes and proplatelets [22]. CD34+ hematopoietic stem/progenitor cells obtained from human cord blood and differentiated to MKs and proplatelets. At day 14 of culture, MKs show proplatelet formation (pointed with red arrows). The color code of the fluorescence staining is as follows: DNA (DAPI, purple-blue), α-tubulin (green), and actin (red).

megakaryopoiesis, which will eminently translate into a regulation, a priori quantitative, of platelet formation.

In this part, we discuss those miRNAs that have been described as regulatory entities for the expression of genes encoding proteins of the Janus kinase (JAK)-signal transducer and activator of transcription (STAT) pathway, which is a fundamental pathway that controls platelet homeostasis. Indeed, mutations in those genes of this pathway, such as JAK2V617F in JAK2 and mutations in MPL exon 10, that cause a constitutive activation of the pathway, developing a thrombocytosis phenotype, named essential thrombocythemia (ET), have been described. As stated above, it has been reported that miR-28 targets the 3'UTR region of MPL, inhibiting its translation and probably contributing to reduce the number of MKs forming proplatelets [65], and therefore platelets number. The regulation of JAK2 by miRNAs in the context of thrombopoiesis is unknown, but Navarro et al. have described direct regulation of JAK2 by miR-135a by targeting its 3'UTR region in Hodgkin lymphoma [70]. Interestingly, the same author in another study described that SOCS1 and SOCS3, which are negative regulators of the JAK/STAT pathway, are directly regulated by miR-203 and miR-221 (a highly platelet expressed miRNA), respectively [71]. In addition, the levels of expression of these two miRNAs maintain a highly inverse correlation with the levels of SOCS1 and SOCS3 in platelets from patients with ET, suggesting that, in combination with a epigenetic regulation, those miRNAs could explain the SOCS1 and SOCS3 downregulated state in ET JAK2V617-negative patients, activating JAK/STAT pathway [71].

Finally and besides the JAK/STAT pathway, Rowley et al. observed in a specific MK Dicer-deficient murine model reduced levels in most platelet miRNAs, which lead to an altered profile of mRNA that apart from the functional consequences related to the increase of Itga2b (αIIb) and Itgb3 (β3), also provokes a mild thrombocytopenia [72].

5. miRNAs in platelet physiology and function

In previous sections, we dealt with quantitative changes observed in the production of platelets by miRNA. In contrast, in this section we deal with the qualitative changes that these can produce in platelets. The role of miRNAs in hemostasis is not well known, and most of the studies have focused on platelets as the central elements of primary hemostasis. Indeed, the first study reporting the presence of miRNAs in platelets was performed in 2008 in healthy controls and patients with polycythemia vera [73]. Of note, platelet miRNA levels have been recently correlated with platelet maturity and platelet function in patients with essential thrombocythemia [74]. Afterwards, many studies reported the presence of a large number of miRNAs in platelets [75]. Importantly, Landry et al. published a landmark paper that established miRNAs as important effectors of platelet function [19]. The authors showed that miRNAs were not only inherited from megakaryocytes, but platelets were also able to produce mature miRNAs from pre-miRNA since they contain all the machinery permitting the miRNA maturation (i.e., Dicer, TRBP2, and Ago2). Different circumstances make uneasy the study of miRNA function in platelets and until now their role still remains elusive. The main reasons are that platelets do not perform transcription and in vitro experiments with platelets are difficult to develop. Since it has been shown by different reports that pre-mRNA maturation as well as de novo translation occurs in platelets after agonist activation such as thrombin [76, 77], different studies have investigated if the same happens for miRNAs since, as mentioned above, they contain the necessary molecular tools for this process. The results are controversial since whereas Corduan et al. showed that thrombin does not provoke miRNA maturation [78], other studies have shown that agonists such as thrombin receptor-activating peptide, collagen, and adenosine diphosphate (ADP) regulate miRNA levels [79]; the difference between both studies may reside in the incubation time, 10 min versus 120 min in the latter study. Thus, as suggested by Corduan et al., quantitative changes in miRNA level unlikely modulate de novo protein synthesis in the minutes following platelet activation, the regulation may rather occur during the whole platelet lifetime by mechanisms that are still to be fully characterized [77]. Upon activation, platelets are able to synthesize proteins such as interleukin-1β (IL-1β), plasminogen activator inhibitor-1 (PAI-1), or thrombospondin-1 (TSP-1). Interestingly, Miao et al. found that platelet activation with thrombin for 30 minutes modified the expression of 103 miRNAs [80]. Among these miRNAs, the authors showed that miR-27b levels, that target TSP-1 [81], decreased after thrombin activation, while TSP-1 increased. The authors also showed using a patented platelet transfection reagent (Ribojuice) that transfection of platelets with a miR-27b mimic decreased TSP-1 levels after thrombin activation [80]. Additionally, Corduan et al. suggested that several de novo translated proteins, among them integrins αIIb and β3, may be regulated by platelet-expressed miRNAs [78]. Interestingly, Rowley et al. demonstrated that in a conditional mouse model (MK) of Dicer1 deficiency, which provokes a reduction of certain miRNA expression levels, mice have high levels of platelet αIIbβ3, elevated platelet reactivity, shortened tail-bleeding time, and reduced survival following collagen/epinephrine-induced pulmonary embolism [72]. In particular, the authors suggest that miR-128, miR-326, miR-331, and miR-500 may regulate αIIbβ3 expression (**Table 3**). Recently, Middleton et al. demonstrated that sepsis increased αIIbβ3 translation in a cecal ligation and puncture (CLP) mouse model; it would be of interest to test if miRNAs involved in the regulation of these two proteins are downregulated and may be implicated in αIIbβ3 overexpression [82].

miRNA.	Target	Ref.
miR-223	$P2Y_{12}$ Factor XIII	[19, 83]
miR-126-3p	ADAM9 PLXNB2	[84]
miR-96	VAMP8	[85]
miR-21	WASP	[86]
miR-27b	TSP-1	[80]
miR-30c	PAI-1	[87]
miR-181a	RAP1B	[88]
miR-128 miR-500 miR-331 miR-326	αIIβ αIIβ,β3 αIIβ,β3 αIIβ	[72]
miR-24	αIIβ	[78]
miR-148a	TULA-2	[89]
miR-376c	PC-TP	[22]
miR-26b miR-140	SELP	[90]

Table 3.
miRNA-targets associated with platelets functionality.

Indeed, Szilágyi et al. recently studied the regulation of miRNAs in a CLP model and observed an important dysregulation of miRNAs in platelets that may explain the pro-thrombotic phenotype caused by this pathology.

Besides pathological conditions, genetic factors may also alter miRNA levels. For example, miR-126 rs4636297 SNP drives the efficiency of miR-126-3p transcription and has been associated with several plasma markers of platelet reactivity such as P-selectin (SELP) or PAF4 suggesting that lower miR-126-3p platelet levels would lead to lower platelet reactivity [84]. Zhou et al. recently showed another potential regulatory pathway for miR-126-3p [91]. The authors discovered that platelets from patients with diabetes mellitus (DM) type 2 had higher levels of long noncoding RNA metallothionein 1 pseudogene 3 (MT1P3) than healthy controls. Interestingly, MT1P3 is able to sponge miR-126-3p and provoke an overexpression of $P2Y_{12}$. This mechanism would additionally explain the high levels of ADP receptor observed in DM2 patients.

As indicated previously, investigating the role of miRNAs in platelet is sometimes an arduous task given the fact that basic cellular assays such as transfection or cell culture are difficult to perform. Investigators are trying to circumvent these problems by developing/adapting transfection reagent [80, 92], but also as nicely shown by García et al. by using alternative approaches [93]. In this paper, the authors demonstrated by using platelet-like structures obtained by differentiating human HSC CD34+ that miR-126-3p is involved in platelet function by regulating Plexin B2 (PLXNB2), an actin dynamics regulator [93]. The use of this kind of technology may help to better define the role of miRNAs in platelets.

6. Platelets-derived miRNA as biomarkers in pathophysiological conditions

Interestingly, platelet miRNAs are the major source of the circulating miRNA pool. Therefore, circulating miRNAs may be interesting biomarkers of disease and of diverse pathophysiological conditions [13]. Indeed, platelets have been implicated in pathological processes such as cardiovascular disease (CVD) [94], for example, myocardial infarction (MI), hypertension, stroke, atrial fibrillation, thrombosis or atherosclerosis, and another pathologies, for example, sepsis or diabetes. Platelet activation plays an important role in the development of the different pathological cardiovascular situations mentioned above. On the other hand, in situations such as sepsis, platelet activation may be a consequence of the disease. In one way or another, platelet activation is essential in these pathologies and miRNAs could be used as a molecular biomarker of this process [13].

miRNAs, whose expression in platelets is consensually high, have been described as independent biomarkers in CVD. For example, let-7 g and miR-191 are independent biomarkers of chronic kidney disease among patients with hypertension [95]. Also, miR-22 and miR-223 were reduced in hypertensive patients with cardiovascular complication being negatively correlated with systolic blood pressure. Furthermore, miR-126 levels were indicative of cardiovascular disease in this patient cohort [96]. miR-21 has been demonstrated to inhibit inflammatory responses in the early phase of MI by targeting KBTBD7 and impairing MKK3/6 activation in immune cells, which subsequently prevented excessive scar formation and improving cardiac function in mice [97].

A special mention should be made of miR-223, abundantly expressed in platelets and highly detected in plasma. This miRNA has been involved in the pathogenesis of different cardiovascular diseases through its effect in platelet reactivity and in endothelial cells [98, 99]. Furthermore, miR-223 may become a reporter of the efficacy of anti-platelet therapy, since it was described that circulating miR-223 may serve as a novel biomarker to assess clopidogrel responsiveness in troponin-negative non-ST elevation acute coronary syndrome patients [100]. Therefore, platelet-derived miRNAs have been related with different cardiovascular diseases and might have important roles as biomarkers not only for cardiovascular disease susceptibility, but also for its prognosis and treatment.

Finally, and as mentioned above, platelets-derived miRNAs can also be helpful from the point of view of biomarkers in other pathological situations such as sepsis or diabetes. Little is known about the regulation of miRNAs in platelets; conditions such as sepsis may play an important role. Indeed, septic platelets showed an altered miRNA profile and reduced platelet miR-26b correlated with sepsis severity and mortality [101]. Therefore, it could become a useful biomarker to indicate the high state of platelet activation in this disease. Other pathological conditions may also alter miRNA levels and impact platelet function. In diabetes, calpain, which is increased in platelets, may cleave Dicer, decreasing the levels of several miRNAs such as miR-223 that may affect functional targets such as the purinergic ADP receptor P2Y12, ultimately promoting the development of thrombosis [19, 83]. Other miRNAs were also found to be altered in diabetes. Fejes et al. showed that miR-26b levels were reduced in DM2 patients with an impact in the levels of SELP [90].

7. Involvement of platelets in intercellular communication

EVs were discovered by Peter Wolf in 1967 and named as "platelets dust" [102]. EVs are basically divided into exosomes, having the smallest diameter size from 30 to 100 nm; apoptotic bodies, which are larger in size (1–5 μm), released during apoptosis; and microvesicles or microparticles ranging in size from 0.1 to 1 μm. Microparticles are cell-derived vesicles that lack the synthetic capacity and may contain proteins, mRNAs, miRNAs, and cytoskeletal components. Platelet-derived microparticles (PMPs) are the most abundant microparticles in circulation and they can be produced from platelet itself or from megakaryocytes. CD42+, CD41+ CD61+/CD31− are some surface markers to identify PMPs, and phosphatidylserine exposure in the surface; whereas CD63 is a marker for platelet exosomes [103]. PMPs have different functions and can be released spontaneously depending on αIIβ3 integrin and cytoskeletal turnover. Platelets activation enhances MPs shedding, and depending on the stimulus or stress, differences in MPs cargo or number have been characterized. Indeed, increased PMPs have been described in myocardial infarction, hypertension, thrombosis, sepsis, diabetes, and other pathologies. Additionally, the platelet activation with different agonists showed that miR-223 was the most abundant miRNA in all samples, and 46 miRNAs were common to all conditions. An enrichment of these 46 miRNAs (in particular, miR-451a and miR-21) was observed in PMPs compared with platelet cargo. Interestingly, platelet activation by CRP (GPVI agonist) generated fourfold more vesicles than the stimulation with ADP, PAR1, or PAR4. Furthermore, platelets stimulated with thrombin shed MPs containing AGO2 and miR-223 complexes that were internalized by endothelial cells (ECs) targeting some genes such as FBXW and EFNA1 [104]. Indeed, miR-223 from PMPs regulated ECs apoptosis targeting insulin-like growth factor 1 receptor [105]. These MPs were also internalized by ECs progenitors in culture [106]. Importantly, PMPs are not only uptaken by endothelial cells. These microparticles play different roles in coagulation and thrombosis, immune response and endothelial senescence and permeability. Thus, PMPs containing miR-126-3p were internalized by macrophages improving their phagocytic capacity, and regulating mRNAs of cytokines or chemokines [107]. Platelets also have a role beyond hemostasis like cancer. Hence, platelets and PMPs are important in tumor progression. MiR-223 from PMPs increased in lung cancer patients promoting its invasion [108]. However, Michael et al. have described that PMPs were infiltrated in the solid tumor promoting tumor cells apoptosis through inhibition of mitochondrial function by miR-24, one of the most abundant miRNAs in these MPs [109].

Platelets-derived exosomes also have an important role in miRNA transference. Plasma exosomes enriched in miR-223, miR-339, and miR-21 were shed before thrombosis in a stenosis murine model, resulting also increased in platelets from a murine stenosis model before thrombosis. These miRNAs were transferred to vascular smooth muscle cells (VSMCs) inhibiting PDGFRβ, involved in cellular proliferation [110]. The inhibition of PDGFRβ was observed in vivo as well [111]. Nevertheless, Zeng and coworkers found that activated platelet internalization by VSMCs rather than MPs increased miR-223 levels promoting VSMCs differentiation. Platelets from diabetic mice had reduced miR-223 levels and consequently increased VSMCs hyperplasia [112].

The phenomenon of horizontal miRNA transference by platelets and microvesicles internalization by ECs, macrophages, VSMCs, cancer cells, and other cells has been described in hepatocytes regulating their proliferation [113]. Activated platelets due to myocardial infarction lose the expression of specific miRNAs. Thus, some of these miRNAs were transferred to ECs in a microvesicle-dependent manner. In particular, miR-320b was transferred to ECs regulating the expression of ICAM-1 modulating adhesion to their surface [114]. Therefore, platelet miRNA transference has a pivotal role in the influence of intercellular signaling and gene expression reprogramming of different cells and in different pathophysiological situations. Moreover, we cannot exclude the emerging role of other noncoding RNAs such as YRNAs, long noncoding RNAs, and circular RNAs in platelets and in the intercellular communication [115] (**Figure 3**).

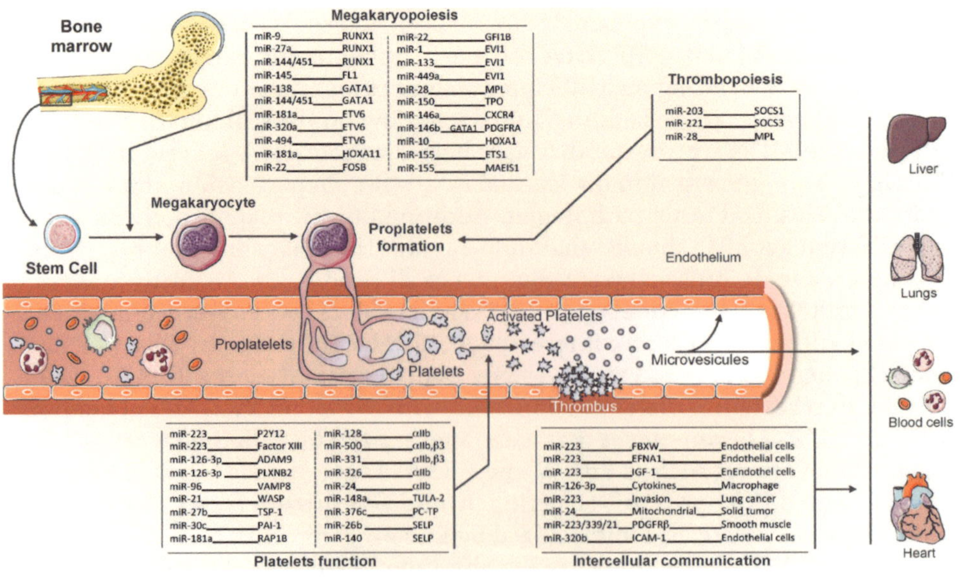

Figure 3.
Graphical abstract.

8. Conclusions

Collectively, the data presented in this chapter indicate that miRNAs have a pivotal role in all platelet states, from their generation regulating megakaryocytes maturation, mainly through transcription factors to their function influencing platelet response under activation (**Figure 3**). Interestingly, miRNAs are important not only for platelet biology, but also because they reflect the platelet status. Thus, platelet miRNAs are used as biomarkers for pathological situations or anti-platelet therapy because they represent the major contribution to circulating miRNA pool. Moreover, miRNAs from platelets may regulate the function of others cells (VSMC, ECs, immune cells, or tumor cells) by intercellular communication mainly through microparticles. Thus, platelet miRNAs have a great potential for future research as therapeutic tools and as biomarkers of disease. Additionally, their role as modulators of platelet function still has to be fully investigated to confirm that their presence in platelets is not a mere inheritance from MKs with any real impact in platelet biology.

Acknowledgements

This work was supported by Instituto de Salud Carlos III and Fondo Europeo de Desarrollo Regional (FEDER; grant no. PI18/0316; PI17/0051) and Fundación Séneca (19873/GERM/15; 20644/JLI/18).

Conflict of interest

The authors declare no conflict of interest.

Author details

Sonia Águila, Ernesto Cuenca-Zamora, Constantino Martínez and Raúl Teruel-Montoya*
Hematology and Medical Oncology Service, HGUJM Morales Meseguer, IMIB-Arrixaca, Transfusion Center, Murcia, Spain

*Address all correspondence to: raulteruelmontoya@hotmail.com

References

[1] Rezaeian A-H, Khanbabaei H, Calin GA. Therapeutic potential of the miRNA-ATM Axis in the management of tumor radioresistance. Cancer Research. 2020;**80**(2):139-150

[2] Gregory TR. Synergy between sequence and size in large-scale genomics. Nature Reviews. Genetics. 2005;**6**(9):699-708

[3] Shu J, Vieira B, Gao T, Xu Z, Cui J. Dynamic and modularized microRNA regulation and its implication in human cancers. Scientific Reports. 2017;7:13356. DOI: 10.1038/s41598-017-13470-5

[4] Lekka E, Hall J. Noncoding RNAs in disease. FEBS Letters. 2018 Sep;**592**(17):2884-2900

[5] Catalanotto C, Cogoni C, Zardo G. MicroRNA in control of gene expression: An overview of nuclear functions. International Journal of Molecular Sciences. 2016;**17**(10):1712

[6] Wienholds E, Kloosterman WP, Miska E, Alvarez-Saavedra E, Berezikov E, de Bruijn E, et al. MicroRNA expression in zebrafish embryonic development. Science. 2005;**309**(5732):310-311

[7] Jackson RJ, Standart N. How do microRNAs regulate gene expression? Science's STKE. 2007;**367**:re1

[8] Alsayegh K, Cortés-Medina LV, Ramos-Mandujano G, Badraiq H, Li M. Hematopoietic differentiation of human pluripotent stem cells: HOX and GATA transcription factors as master regulators. Current Genomics. 2019;**20**(6):438-452

[9] Kim M, Civin CI, Kingsbury TJ. MicroRNAs as regulators and effectors of hematopoietic transcription factors. Wiley Interdisciplinary Reviews: RNA. 2019;**10**(5):e1537

[10] Mittelbrunn M, Sánchez-Madrid F. Intercellular communication: diverse structures for exchange of genetic information. Nature Reviews. Molecular Cell Biology. 2012;**13**:328-335

[11] Teruel-Montoya R, Luengo-Gil G, Vallejo F, Yuste JE, Bohdan N, García-Barberá N, et al. Differential miRNA expression profile and proteome in plasma exosomes from patients with paroxysmal nocturnal hemoglobinuria. Scientific Reports. 2019;**9**(1):3611

[12] Teruel-montoya R, Kong X, Abraham S, Ma L, Kunapuli SP, Holinstat M, et al. MicroRNA expression differences in human hematopoietic cell lineages enable regulated transgene expression. PLoS One. 2014;**9**(7):1-12

[13] Willeit P, Zampetaki A, Dudek K, Kaudewitz D, King A, Kirkby NS, et al. Circulating microRNAs as novel biomarkers for platelet activation. Circulation Research. 2013;**112**(4):595-600

[14] Berckmans RJ, Nieuwland R, Böing AN, Romijn FP, Hack CE, Sturk A. Cell-derived microparticles circulate in healthy humans and support low grade thrombin generation. Thrombosis and Haemostasis. 2001;**85**(4):639-646

[15] Denli AM, Tops BBJ, Plasterk RHA, Ketting RF, Hannon GJ. Processing of primary microRNAs by the microprocessor complex. Nature. 2004;**432**(7014):231-235

[16] Liu Z, Wang J, Cheng H, Ke X, Sun L, Zhang QC, et al. Cryo-EM structure of human dicer and its complexes with a pre-miRNA substrate. Cell. 2018;**173**(5):1191-1203.e12

[17] Jin W, Wang J, Liu C-P, Wang H-W, Xu R-M. Structural basis for pri-miRNA

recognition by drosha. Molecular Cell. 2020;**78**:1-11

[18] Gebert LFR, MacRae IJ. Regulation of microRNA function in animals. Nature Reviews. Molecular Cell Biology. 2019;**20**(1):21-37

[19] Landry P, Plante I, Ouellet DL, Perron MP, Rousseau G, Provost P. Existence of a microRNA pathway in anucleate platelets. Nature Structural & Molecular Biology. 2009;**16**(9):961-966

[20] Chatterjee A, Leichter AL, Fan V, Tsai P, Purcell RV, Sullivan MJ, et al. A cross comparison of technologies for the detection of microRNAs in clinical FFPE samples of hepatoblastoma patients. Scientific Reports. 2015;**5**:10438

[21] Nagalla S, Shaw C, Kong X, Kondkar AA, Edelstein LC, Ma L, et al. Platelet microRNA-mRNA coexpression profiles correlate with platelet reactivity. Blood. 2011;**117**(19):5189-5197

[22] Edelstein LC, Simon LM, Montoya RT, Holinstat M, Chen ES, Bergeron A, et al. Racial differences in human platelet PAR4 reactivity reflect expression of PCTP and miR-376c. Nature Medicine. 2013;**19**(12):1609-1616

[23] Simon LM, Edelstein LC, Nagalla S, Woodley AB, Chen ES, Kong X, et al. Human platelet microRNA-mRNA networks associated with age and gender revealed by integrated plateletomics. Blood. 2014;**123**(16):e37-e45

[24] Teruel-Montoya R, Kong X, Abraham S, Ma L, Kunapuli SP, Holinstat M, et al. MicroRNA expression differences in human hematopoietic cell lineages enable regulated transgene expression. PLoS One. 2014;**9**(7):e102259

[25] Mitchell PS, Parkin RK, Kroh EM, Fritz BR, Wyman SK, Pogosova-Agadjanyan EL, et al. Circulating microRNAs as stable blood-based markers for cancer detection. Proceedings of the National Academy of Sciences of the United States of America. 2008;**105**(30):10513-10518

[26] Chen X, Ba Y, Ma L, Cai X, Yin Y, Wang K, et al. Characterization of microRNAs in serum: A novel class of biomarkers for diagnosis of cancer and other diseases. Cell Research. 2008;**18**(10):997-1006. Available from: http://www.nature.com/doifinder/10.1038/cr.2008.282

[27] Machlus KR, Italiano JEJ. The incredible journey: From megakaryocyte development to platelet formation. The Journal of Cell Biology. 2013;**201**(6):785-796

[28] Kohrs N, Kolodziej S, Kuvardina ON, Herglotz J, Yillah J, Herkt S, et al. MiR144/451 Expression Is Repressed by RUNX1 During Megakaryopoiesis and Disturbed by RUNX1/ETO. PLoS Genetics. 2016;**12**(3):e1005946

[29] Lordier L, Bluteau D, Jalil A, Legrand C, Pan J, Rameau P, et al. RUNX1-induced silencing of non-muscle myosin heavy chain IIB contributes to megakaryocyte polyploidization. Nature Communications. 2012;**3**:717

[30] Bluteau D, Glembotsky AC, Raimbault A, Balayn N, Gilles L, Rameau P, et al. Dysmegakaryopoiesis of FPD/AML pedigrees with constitutional RUNX1 mutations is linked to myosin II deregulated expression. Blood. 2012;**120**(13):2708-2718

[31] Dore LC, Amigo JD, dos Santos CO, Zhang Z, Gai X, Tobias JW, et al. A GATA-1-regulated microRNA locus essential for erythropoiesis. Proceedings of the National Academy of Sciences. 2008;**105**(9):3333-3338

[32] Zhang L, Sankaran VG, Lodish HF. MicroRNAs in erythroid and megakaryocytic differentiation and megakaryocyte–erythroid progenitor lineage commitment. Leukemia. 2012;**26**(11):2310-2316

[33] Rasmussen KD, Simmini S, Abreu-Goodger C, Bartonicek N, Di Giacomo M, Bilbao-Cortes D, et al. The miR-144/451 locus is required for erythroid homeostasis. The Journal of Experimental Medicine. 2010;**207**(7):1351-1358

[34] Ben-ami O, Pencovich N, Lotem J, Levanon D, Groner Y. A regulatory interplay between miR-27a and Runx1 during megakaryopoiesis. Proceedings of the National Academy of Sciences of the United States of America USA. 2009;**106**:238-243

[35] Raghuwanshi S, Gutti U, Kandi R, Gutti RK. MicroRNA-9 promotes cell proliferation by regulating RUNX1 expression in human megakaryocyte development. Cell Proliferation. 2018;**51**:e12414

[36] Bluteau O, Langlois T, Favale F, Rameau P. Developmental changes in human megakaryopoiesis. Journal of Thrombosis and Haemostasis. 2013;**11**:1730-1741

[37] Labbaye C, Spinello I, Quaranta MT, Pelosi E, Pasquini L, Petrucci E, et al. A three-step pathway comprising PLZF/miR-146a/CXCR4 controls megakaryopoiesis. Nature Cell Biology. 2008;**10**(7):788-801

[38] Lulli V, Romania P, Morsilli O, Gabbianelli M, Pagliuca A, Mazzeo S, et al. Overexpression of Ets-1 in human hematopoietic progenitor cells blocks erythroid and promotes megakaryocytic differentiation. Cell Death and Differentiation. 2006;**13**(7):1064-1074

[39] Wang H, Liu C, Liu X, Wang M, Wu D, Gao J, et al. MEIS1 regulates hemogenic endothelial generation, megakaryopoiesis, and thrombopoiesis in human pluripotent stem cells by targeting TAL1 and FLI1. Stem Cell Reports. 2018;**10**(2):447-460

[40] Romania P, Lulli V, Pelosi E, Biffoni M, Peschle C, Marziali G. MicroRNA 155 modulates megakaryopoiesis at progenitor and precursor level by targeting Ets-1 and Meis1 transcription factors. British Journal of Haematology. 2008;**143**(4):570-580

[41] Opalinska JB, Bersenev A, Zhang Z, Schmaier AA, Choi J, Yao Y, et al. MicroRNA expression in maturing murine megakaryocytes. Blood. 2010;**116**(23):128-138

[42] Raghuwanshi S, Dahariya S, Musvi SS, Gutti U, Kandi R, Undi RB, et al. MicroRNA function in megakaryocytes. Platelets. 2019;**30**(7):809-816

[43] Kumar MS, Narla A, Nonami A, Mullally A, Dimitrova N, Ball B, et al. Coordinate loss of a microRNA and protein-coding gene cooperate in the pathogenesis of 5q− syndrome. Blood. 2011;**118**(17):4666-4673

[44] Liu T, Xia L, Yao Y, Yan C, Fan Y, Gajendran B, et al. Identification of diterpenoid compounds that interfere with Fli-1 DNA binding to suppress leukemogenesis. Cell Death & Disease. 2019;**10**(2):117

[45] Xu C, Fu H, Gao L, Wang L, Wang W, Li J, et al. BCR-ABL/GATA1/miR-138 mini circuitry contributes to the leukemogenesis of chronic myeloid leukemia. Oncogene. 2014;**33**(1):44-54

[46] Undi RB, Kandi R, Gutti RK. MicroRNAs as haematopoiesis regulators. Advances in Hematology. 2013;**2013**:1-20

[47] Garzon R, Pichiorri F, Palumbo T, Iuliano R, Cimmino A, Aqeilan R, et al. MicroRNA fingerprints during human

megakaryocytopoiesis. Proceedings of the National Academy of Sciences. 2006;**103**(13):5078-5083

[48] Zarif MN, Soleimani M, Abolghasemi H, Amirizade N, Arefian E, Rahimian A. Megakaryocytic differentiation of CD133+ hematopoietic stem cells by down-regulation of microRNA-10a. Hematology. 2013;**18**(2):93-100

[49] Zhou S-F, He S-M, Zeng S, Zhou Z-W, He Z. Hsa-microRNA-181a is a regulator of a number of cancer genes and a biomarker for endometrial carcinoma in patients: A bioinformatic and clinical study and the therapeutic implication. Drug Design, Development and Therapy. 2015;**9**:1103

[50] Yang Y-L, Yen C-T, Pai C-H, Chen H-Y, Yu S-L, Lin C-Y, et al. A double negative loop comprising ETV6/RUNX1 and MIR181A1 contributes to differentiation block in t(12,21)-positive acute lymphoblastic leukemia. Lausen J, editor. PLoS One. 2015;**10**(11):e0142863

[51] Diakos C, Zhong S, Xiao Y, Zhou M, Vasconcelos GM, Krapf G, et al. TEL-AML1 regulation of survivin and apoptosis via miRNA-494 and miRNA-320a. Blood. 2010;**116**(23):4885-4893

[52] Möröy T, Vassen L, Wilkes B, Khandanpour C. From cytopenia to leukemia: The role of Gfi1 and Gfi1b in blood formation. Blood. 2015;**126**(24):2561-2569

[53] Weiss CN, Ito K. MicroRNA-22 promotes megakaryocyte differentiation through repression of its target, GFI1. Blood Advances. 2019;**3**(1):33-46

[54] Ferrer-Marín F, Arroyo AB, Bellosillo B, Cuenca EJ, Zamora L, Hernández-Rivas JM, et al. miR-146a rs2431697 identifies myeloproliferative neoplasm patients with higher secondary myelofibrosis progression risk. Leukemia. 2020. DOI: 10.1038/s41375-020-0767-3

[55] Starczynowski DT, Kuchenbauer F, Argiropoulos B, Sung S, Morin R, Muranyi A, et al. Identification of miR-145 and miR-146a as mediators of the 5q– syndrome phenotype. Nature Medicine. 2010;**16**(1):49-58

[56] Starczynowski DT, Kuchenbauer F, Wegrzyn J, Rouhi A, Petriv O, Hansen CL, et al. MicroRNA-146a disrupts hematopoietic differentiation and survival. Experimental Hematology. 2011;**39**(2):167-178.e4

[57] Limb J-K, Yoon S, Lee KE, Kim B-H, Lee S, Bae YS, et al. Regulation of megakaryocytic differentiation of K562 cells by FosB, a member of the Fos family of AP-1 transcription factors. Cellular and Molecular Life Sciences. 2009;**66**(11-12):1962-1973

[58] Kim M, Civin CI, Kingsbury TJ. MicroRNAs as regulators and effectors of hematopoietic transcription factors. Wiley Interdisciplinary Reviews: RNA. 2019;**10**(5):e1537

[59] Asri A, Sabour J, Atashi A, Soleimani M. Homing in hematopoietic stem cells: Focus on regulatory role of CXCR7 on SDF1A/CXCR4 axis. EXCLI Journal. 2016;**15**:134-143

[60] Pang L, Weiss MJ, Poncz M. Megakaryocyte biology and related disorders. The Journal of Clinical Investigation. 2005;**115**(12):3332-3338

[61] Gómez-Benito M, Conchillo A, García MA, Vázquez I, Maicas M, Vicente C, et al. EVI1 controls proliferation in acute myeloid leukaemia through modulation of miR-1-2. British Journal of Cancer. 2010;**103**(8):1292-1296

[62] Yamamoto H, Lu J, Oba S, Kawamata T, Yoshimi A, Kurosaki N, et al. miR-133 regulates Evi1 expression in AML cells as

a potential therapeutic target. Scientific Reports. 2016;**6**(1):19204

[63] De Weer A, Van der Meulen J, Rondou P, Taghon T, Konrad TA, De Preter K, et al. EVI1-mediated down regulation of MIR449A is essential for the survival of EVI1 positive leukaemic cells. British Journal of Haematology. 2011;**154**(3):337-348

[64] Zhai P-F, Wang F, Su R, Lin H-S, Jiang C-L, Yang G-H, et al. The regulatory roles of microRNA-146b-5p and its target platelet-derived growth factor receptor α (PDGFRA) in erythropoiesis and megakaryocytopoiesis. The Journal of Biological Chemistry. 2014;**289**(33):22600-22613

[65] Girardot M, Pecquet C, Boukour S, Knoops L, Ferrant A, Vainchenker W, et al. miR-28 is a thrombopoietin receptor targeting microRNA detected in a fraction of myeloproliferative neoplasm patient platelets. Blood. 2010;**116**(3):437-445

[66] Barroga CF, Pham H, Kaushansky K. Thrombopoietin regulates c-Myb expression by modulating micro RNA 150 expression. Experimental Hematology. 2008;**36**(12):1585-1592

[67] Lu J, Guo S, Ebert BL, Zhang H, Peng X, Bosco J, et al. MicroRNA-mediated control of cell fate in megakaryocyte-erythrocyte progenitors. Developmental Cell. 2008;**14**(6):843-853

[68] Cuenca-zamora EJ, Ferrer-mar F, Rivera J. Tubulin in platelets: When the shape matters. International Journal of Molecular Sciences. 2019; **20**(14):3484

[69] Songdej N, Rao AK. Hematopoietic transcription factor mutations: Important players in inherited platelet defects. Blood. 2017;**129**(21):2873-2881

[70] Navarro A, Diaz T, Martinez A, Gaya A, Pons A, Gel B, et al. Regulation of JAK2 by miR-135a: Prognostic impact in classic Hodgkin lymphoma. Blood. 2009;**114**(14):2945-2951

[71] Navarro A, Pairet S, Álvarez-Larrán A, Pons A, Ferrer G, Longarón R, et al. miR-203 and miR-221 regulate SOCS1 and SOCS3 in essential thrombocythemia. Blood Cancer Journal. 2016;**6**(3):e406

[72] Rowley JW, Chappaz S, Corduan A, Chong MMW, Campbell R, Khoury A, et al. Dicer1-mediated miRNA processing shapes the mRNA profile and function of murine platelets. Blood. 2016;**127**(14):1743-1751

[73] Bruchova H, Merkerova M, Prchal JT. Aberrant expression of microRNA in polycythemia vera. Haematologica. 2008;**93**(7):1009-1016

[74] Tran JQD, Pedersen OH, Larsen ML, Grove EL, Kristensen SD, Hvas A-M, et al. Platelet microRNA expression and association with platelet maturity and function in patients with essential thrombocythemia. Platelets. 2020;**31**(3):365-372

[75] Plé H, Landry P, Benham A, Coarfa C, Gunaratne PH, Provost P. The repertoire and features of human platelet microRNAs. PLoS One. 2012;**7**(12):e50746

[76] Denis MM, Tolley ND, Bunting M, Schwertz H, Jiang H, Lindemann S, et al. Escaping the nuclear confines: signal-dependent pre-mRNA splicing in anucleate platelets. Cell. 2005;**122**(3):379-391

[77] Weyrich AS, Schwertz H, Kraiss LW, Zimmerman GA. Protein synthesis by platelets: Historical and new perspectives. Journal of Thrombosis and Haemostasis. 2009;**7**(2):241-246

[78] Corduan A, Plé H, Laffont B, Wallon T, Plante I, Landry P, et al. Dissociation of SERPINE1 mRNA from the translational repressor proteins Ago2 and TIA-1 upon platelet activation. Thrombosis and Haemostasis. 2015;**113**(5):1046-1059

[79] Cimmino G, Tarallo R, Nassa G, De Filippo MR, Giurato G, Ravo M, et al. Activating stimuli induce platelet microRNA modulation and proteome reorganisation. Thrombosis and Haemostasis. 2015;**114**(1):96-108

[80] Miao X, Rahman MF-U, Jiang L, Min Y, Tan S, Xie H, et al. Thrombin-reduced miR-27b attenuates platelet angiogenic activities in vitro via enhancing platelet synthesis of anti-angiogenic thrombospondin-1. Journal of Thrombosis and Haemostasis. 2018;**16**(4):791-801

[81] Wang J-M, Tao J, Chen D-D, Cai J-J, Irani K, Wang Q, et al. MicroRNA miR-27b rescues bone marrow-derived angiogenic cell function and accelerates wound healing in type 2 diabetes mellitus. Arteriosclerosis, Thrombosis, and Vascular Biology. 2014;**34**(1):99-109

[82] Middleton EA, Rowley JW, Campbell RA, Grissom CK, Brown SM, Beesley SJ, et al. Sepsis alters the transcriptional and translational landscape of human and murine platelets. Blood. 2019;**134**(12):911-923

[83] Elgheznawy A, Shi L, Hu J, Wittig I, Laban H, Pircher J, et al. Dicer cleavage by calpain determines platelet microRNA levels and function in diabetes. Circulation Research. 2015;**117**(2):157-165

[84] Kaudewitz D, Skroblin P, Bender LH, Barwari T, Willeit P, Pechlaner R, et al. Association of MicroRNAs and YRNAs with platelet function. Circulation Research. 2016;**118**(3):420-432

[85] Kondkar AA, Bray MS, Leal SM, Nagalla S, Liu DJ, Jin Y, et al. VAMP8/endobrevin is overexpressed in hyperreactive human platelets: Suggested role for platelet microRNA. Journal of Thrombosis and Haemostasis. 2010;**8**(2):369-378

[86] Barwari T, Eminaga S, Mayr U, Lu R, Armstrong PC, Chan MV, et al. Inhibition of profibrotic microRNA-21 affects platelets and their releasate. JCI Insight. 2018;**3**(21):e123335

[87] Luo M, Li R, Ren M, Chen N, Deng X, Tan X, et al. Hyperglycaemia-induced reciprocal changes in miR-30c and PAI-1 expression in platelets. Scientific Reports. 2016;**6**:36687

[88] Dahiya N, Atreya CD. MiR-181a reduces platelet activation via the inhibition of endogenous RAP1B. MicroRNA. 2020;**9**(3):240-246

[89] Zhou Y, Abraham S, Andre P, Edelstein LC, Shaw CA, Dangelmaier CA, et al. Anti-miR-148a regulates platelet FcγRIIA signaling and decreases thrombosis in vivo in mice. Blood. 2015;**126**(26):2871-2881

[90] Fejes Z, Póliska S, Czimmerer Z, Káplár M, Penyige A, Gál Szabó G, et al. Hyperglycaemia suppresses microRNA expression in platelets to increase P2RY12 and SELP levels in type 2 diabetes mellitus. Thrombosis and Haemostasis. 2017;**117**(3):529-542

[91] Zhou M, Gao M, Luo Y, Gui R, Ji H. Long non-coding RNA metallothionein 1 pseudogene 3 promotes p2y12 expression by sponging miR-126 to activate platelet in diabetic animal model. Platelets. 2019;**30**(4):452-459

[92] Hong W, Kondkar AA, Nagalla S, Bergmeier W, Jin Y, Herman JH, et al. Transfection of human platelets with short interfering RNA. Clinical and Translational Science. 2011;**4**(3):180-182

[93] Garcia A, Dunoyer-Geindre S, Zapilko V, Nolli S, Reny J-L, Fontana P. Functional validation of microRNA-126-3p as a platelet reactivity regulator using human haematopoietic stem cells. Thrombosis and Haemostasis. 2019;**119**(2):254-263

[94] McManus DD, Freedman JE. MicroRNAs in platelet function and cardiovascular disease. Nature Reviews Cardiology. 2015;**12**(12):711-717. Available from: http://www.nature.com/doifinder/10.1038/nrcardio.2015.101

[95] Berillo O, Huo K-G, Fraulob-Aquino JC, Richer C, Briet M, Boutouyrie P, et al. Circulating let-7g-5p and miR-191-5p are independent predictors of chronic kidney disease in hypertensive patients. American Journal of Hypertension. 2020;**33**(6):505-513

[96] Marketou M, Kontaraki J, Papadakis J, Kochiadakis G, Vrentzos G, Maragkoudakis S, et al. Platelet microRNAs in hypertensive patients with and without cardiovascular disease. Journal of Human Hypertension. 2019;**33**(2):149-156

[97] Yang L, Wang B, Zhou Q, Wang Y, Liu X, Liu Z, et al. MicroRNA-21 prevents excessive inflammation and cardiac dysfunction after myocardial infarction through targeting KBTBD7. Cell Death & Disease. 2018;**9**(7):769

[98] Chu M, Wu R, Qin S, Hua W, Shan Z, Rong X, et al. Bone marrow-derived microRNA-223 works as an endocrine genetic signal in vascular endothelial cells and participates in vascular injury from Kawasaki disease. Journal of the American Heart Association. 2017;**6**(2):e004878

[99] Chyrchel B, Totoń-Żurańska J, Kruszelnicka O, Chyrchel M, Mielecki W, Kołton-Wróż M, et al. Association of plasma miR-223 and platelet reactivity in patients with coronary artery disease on dual antiplatelet therapy: A preliminary report. Platelets. 2015;**26**(6):593-597

[100] Zhang Y-Y, Zhou X, Ji W-J, Shi R, Lu R-Y, Li J-L, et al. Decreased circulating microRNA-223 level predicts high on-treatment platelet reactivity in patients with troponin-negative non-ST elevation acute coronary syndrome. Journal of Thrombosis and Thrombolysis. 2014;**38**(1):65-72

[101] Szilágyi B, Fejes Z, Póliska S, Pócsi M, Czimmerer Z, Patsalos A, et al. Reduced miR-26b expression in megakaryocytes and platelets contributes to elevated level of platelet activation status in sepsis. International Journal of Molecular Sciences. 2020;**21**(3):866

[102] Wolf P. The nature and significance of platelet products in human plasma. British Journal of Haematology. 1967;**13**(3):269-288

[103] Heijnen HF, Schiel AE, Fijnheer R, Geuze HJ, Sixma JJ. Activated platelets release two types of membrane vesicles: Microvesicles by surface shedding and exosomes derived from exocytosis of multivesicular bodies and alpha-granules. Blood. 1999;**94**(11):3791-3799

[104] Laffont B, Corduan A, Plé H, Duchez A-C, Cloutier N, Boilard E, et al. Activated platelets can deliver mRNA regulatory Ago2·microRNA complexes to endothelial cells via microparticles. Blood. 2013;**122**(2):253-261

[105] Pan Y, Liang H, Liu H, Li D, Chen X, Li L, et al. Platelet-secreted microRNA-223 promotes endothelial cell apoptosis induced by advanced glycation end products via targeting the insulin-like growth factor 1 receptor. Journal of Immunology. 2014;**192**(1):437-446

[106] Prokopi M, Pula G, Mayr U, Devue C, Gallagher J, Xiao Q, et al. Proteomic analysis reveals presence of platelet microparticles in endothelial progenitor cell cultures. Blood. 2009;**114**(3):723-732

[107] Laffont B, Corduan A, Rousseau M, Duchez A-C, Lee CHC, Boilard E, et al. Platelet microparticles reprogram macrophage gene expression and function. Thrombosis and Haemostasis. 2016;**115**(2):311-323

[108] Liang H, Yan X, Pan Y, Wang Y, Wang N, Li L, et al. MicroRNA-223 delivered by platelet-derived microvesicles promotes lung cancer cell invasion via targeting tumor suppressor EPB41L3. Molecular Cancer. 2015;**14**:58

[109] Michael JV, Wurtzel JGT, Mao GF, Rao AK, Kolpakov MA, Sabri A, et al. Platelet microparticles infiltrating solid tumors transfer miRNAs that suppress tumor growth. Blood. 2017;**130**(5):567-580

[110] Tan M, Yan H-B, Li J-N, Li W-K, Fu Y-Y, Chen W, et al. Thrombin stimulated platelet-derived exosomes inhibit platelet-derived growth factor receptor-beta expression in vascular smooth muscle cells. International Journal of Experimental Cellular Physiology, Biochemistry and Pharmacology. 2016;**38**(6):2348-2365

[111] Silverman-Gavrila R, Silverman-Gavrila L, Bendeck MP. Cell division fidelity is altered during the vascular response to injury: its novel role in atherosclerosis progression. The American Journal of Pathology. 2013;**182**(3):628-639

[112] Zeng Z, Xia L, Fan X, Ostriker AC, Yarovinsky T, Su M, et al. Platelet-derived miR-223 promotes a phenotypic switch in arterial injury repair. The Journal of Clinical Investigation. 2019;**129**(3):1372-1386

[113] Kirschbaum M, Karimian G, Adelmeijer J, Giepmans BNG, Porte RJ, Lisman T. Horizontal RNA transfer mediates platelet-induced hepatocyte proliferation. Blood. 2015;**126**(6):798-806

[114] Gidlöf O, van der Brug M, Ohman J, Gilje P, Olde B, Wahlestedt C, et al. Platelets activated during myocardial infarction release functional miRNA, which can be taken up by endothelial cells and regulate ICAM1 expression. Blood. 2013;**121**(19):3908-3917. S1-26

[115] Sunderland N, Skroblin P, Barwari T, Huntley RP, Lu R, Joshi A, et al. MicroRNA biomarkers and platelet reactivity: The clot thickens. Circulation Research. 2017;**120**(2):418-435

Section 2

Platelet Dysfunction

Chapter 5

Bleeding Disorders Associated with Abnormal Platelets: Glanzmann Thrombasthenia and Bernard-Soulier Syndrome

Muhammet Mesut Nezir Engin

Abstract

Platelets, the smallest cells in the blood, are associated with hemostasis, bowel formation, tissue remodeling, and wound healing. Although the prevalence of inherited platelet disorders is not fully known, it is a rare disease group and is encountered in approximately between 10000 and 1000000. Glanzmann thrombasthenia (GT) and Bernard-Soulier syndrome (BSS) are more frequently observed in inherited platelet disorders. In GT, the platelet aggregation stage due to deficiency or dysfunction of the platelet GPIIb/IIIa complex cannot take place. BSS is a platelet adhesion disorder due to the absence or abnormality of GPIb/IX complex on the platelet surface. If there is bleeding after easy bruising, mucous and oral cavities, menorrhagia, tooth extraction, tonsillectomy, or other surgical interventions, inherited platelet dysfunction should be considered if the platelet count is normal while the bleeding time is long. Firstly, other causes should be investigated by making differential diagnosis of GT and BSS. In this chapter, the definition, etiology, historical process, epidemiology, genetic basis, pathophysiology, clinical findings, diagnosis, differential diagnosis, and the follow-up and treatment approach of GT and BSS will be reviewed according to the current medical literature.

Keywords: Glanzmann thrombasthenia, Bernard-Soulier syndrome, thrombocyte function disorder, thrombocyte transfusion, rFVIIa

1. Introduction

Platelets, the smallest cells in the blood, are associated with hemostasis, bowel formation, tissue remodeling, and wound healing. Platelets perform their tasks in ensuring hemostasis in four stages: platelet adhesion, activation of platelet, platelet aggregation, and platelet procoagulant activity. When a damage occurs on the vascular endothelial surface, platelets bind to the collagen, fibronectin, von Willebrand factor, thrombospondin, and fibrinogen in the endothelial substrate with the glycoprotein receptors they carry on their surface. In this way, platelet adhesion takes place. Binding of platelet receptors to their respective ligands causes activation of the platelet. This activation occurs as a result of the change in the cytoskeleton system due to intracellular calcium. By importing the impulse from outside the cell, platelet α-granules secrete their contents. The released ADP causes structural

IntechOpen

	Glanzmann thrombasthenia	Bernard-Soulier syndrome
Genetic mutation	17q21 chromosome. ITGA2B or ITGBS genes	GPIbα, GPIbβ, and GPIX genes
Pathophysiology	Deficiency or dysfunction of the platelet GPIIb/IIIa complex	Deficiency or dysfunction of the platelet Gplb/V/IX complex
Affected platelet function	Aggregation	Adhesion

Table 1.
Comparison of genetic mutation, pathophysiology, and affected platelet function status of Glanzmann thrombasthenia and Bernard-Soulier syndrome.

change in GPIIb/IIIa on the platelet surface. Fibrinogen binds two or more platelets via GPIIb/IIIa receptors that are structurally altered, resulting in platelet aggregation. After aggregation of platelets, platelet plugs are formed at the damage site. Activation of platelets leads to changes in phospholipids on their surface. These phospholipids enable the activation of some clotting factors and perform platelet procoagulant activity [1–6].

The problem in any of the functions of platelets creates a tendency for the primary hemostatic plug not to form and therefore to bleed. Platelet dysfunctions can be hereditary or acquired. Although the prevalence of inherited platelet disorders is not fully known, it is a rare disease group and is frequently encountered in approximately between 10000 and 1000000. Glanzmann thrombasthenia (GT) and Bernard-Soulier syndrome (BSS) are more frequently observed in inherited platelet disorders.

In GT, the platelet aggregation stage due to deficiency or dysfunction of the platelet GPIIb/IIIa complex cannot take place. BSS is a platelet adhesion disorder due to the absence or abnormality of GPIb/IX complex on the platelet surface [1, 7, 8] (**Table 1**).

If there is bleeding after easy bruising, mucous and oral cavities, menorrhagia, tooth extraction, tonsillectomy, or other surgical interventions, inherited platelet dysfunction should be considered if the platelet count is normal while the bleeding time is long. Firstly, other causes should be investigated by making differential diagnosis of GT and BSS [1, 7]. In this chapter, the definition, etiology, historical process, epidemiology, genetic basis, pathophysiology, clinical findings, diagnosis, differential diagnosis, and treatment approach of GT and BSS will be reviewed according to the current medical literature.

2. Glanzmann thrombasthenia

2.1 Definition

GT is an autosomal recessive congenital bleeding disorder characterized by a lack of platelet aggregation due to defect and/or deficiency of α IIbβ 3 integrin. Integrin is a platelet fibrinogen receptor, necessary for platelet aggregation and hemostasis. Patients with this disorder often experience lifelong bleeding episodes involving mucocutaneous membranes [9–13].

2.2 History

This disease was first described by Swiss pediatrician Eduard Glanzmann in 1918 as "hereditary hemorrhagic thrombasthenia." Braunsteiner and Pakesch, on the other hand, reviewed platelet dysfunctions in 1956, after which they identified thrombasthenia as a hereditary disease characterized by normal size platelets that did not spread to the surface and did not support clot retraction. The diagnostic characteristics of GT including the absence of platelet aggregation as a primary feature, were reported in 1964 by Caen et al. has been clearly identified by the classical report on 15 French patients. Those patients without platelet aggregation and no clot retraction were later called type I disease patients and those with absent aggregation but residual clot retraction were called type II disease patients; variant disease was first identified in 1987 [8, 10, 14–16].

2.3 Etiology

GT is an autosomal recessive disease with mutations containing the 17q21 chromosome, especially the ITGA2B or ITGB3 genes. GT results when a patient is homozygous for the same mutation or is a compound heterozygote for different mutations. GT is usually caused by decreased or absent expression of αIIb or β3, abnormalities in protein folding, transport of the integrin subunit causing post-translational defective processing or decreased surface expression, or abnormalities affecting protein function. Other defects change the integrin function by altering the ligand binding pocket (interface between αIIb and β3) that changes the cytoplasmic domain and affects the binding of regulators or locks integrin in active form [8, 9, 12, 13, 17–20].

2.4 Epidemiology

GT has an increasing incidence in populations where marriage between close relatives is an accepted tradition. The prevalence is estimated to be approximately 1:1,000,000 in the general population. Research shows that women are slightly more frequently affected than men. For example, when 177 patients with GT in Paris were examined, 102 (58%) of the patients were shown to be women. In addition, 12 patients were in the USA, 55 were in Israel and Jordan, and 42 were in South India. Some patients may have mild symptoms and are never detected to have GT, so the true prevalence may be higher than reported. Type I is the most common subtype and accounts for about 78% of patients with GT type II and type III (functional variant in receptor) and accounts for about 14% and 8% of cases [8, 9, 12].

2.5 Genetic basis

The ITGA2B and ITGB3 genes are found on chromosomes 17q21.31 and 17q21.32, respectively, and are independently expressed. Due to autosomal recessive inheritance, compound heterozygosity is common, except for selected ethnic groups, where homozygosity is more likely due to kinship. A higher percentage of pathogenic variants occur in ITGA2B compared to ITGB3, which consists of 15 exons with 788 amino acids, probably because this gene is larger with 30 exons encoding 1039 amino acids. There is a constantly updated database on the Internet http://sinaicentral.mssm.edu/intranet/research/glanzmann: currently, it contains a list of 558 mutations that lead to GT. In addition, when the data in the database were examined, it was found that 269 patients had homozygous mutations. This shows us that consanguineous marriage is an important feature in the heredity of this disease. Some researchers described that pathogenic, nonsense missense, and splice site variants are commonly observed and large deletions and duplications are rarely observed. Pathogenic missense variants cause the disruption of subunit biosynthesis megakaryocytes or prevention of the exit of pro-αIIbβ3 complexes from endoplasmic reticulum to Golgi device or the cell surface mature complexes. Most of the genetic variants affect the β-propeller region of αIIb and domains of β3 of the epithelial growth factor [20–23].

2.6 Pathophysiology

The main mechanism in the pathophysiology of GT is the qualitative or quantitative disorder of the autosomal recessive platelet surface receptor of GPIIb/IIIa (ITG αIIbβ3). As a result, it results in erroneous platelet aggregation and reduced clot retraction. ITG αIIbβ3 is a large heterodimeric cell transmembrane receptor consisting of a larger αIIb and a smaller P3 subunit. These subunits were not covalently attached to permit bidirectional signal between the cell membrane and extracellular matrix when initiating intracellular signaling pathways. It contains cytoplasmic and transmembrane domains that act as the junction point for intracellular signal molecules and proteins. The activation of ITG αIIbβ3 is provided by the B3 subunit consisting of large disulfide epidermal growth factor (EGF) domains. Calcium binding sites for complex formation and platelet-platelet adhesion are found on the p-propeller region of the αIIb subunit. The receptor head function consisting of binding fibrinogen, VWF, vitronectin, and fibronectin is necessary for platelet communications by regulation of cell migration, platelet aggregation and adhesion, and the formation of a thrombus [24].

The ITGA2B gene, located on chromosome 17q21.31, encodes the platelet GPαIIb, while the gene encoding the glycoprotein subunit IIIa is found in chromosome ITGB3, 17q21.32. Both subunits are collected from the precursors of the endoplasmic reticulum by further processing in the Golgi apparatus. Nurden et al. examined more closely the p-propeller ectodomain mutations of the αIIb subunit. Nurden et al. concluded that a large series of mutations affecting the β-propeller field interrupts calcium binding and has numerous harmful effects on αIIbβ3 expression and function, and causes different types of GT [21, 25, 26].

Homozygous or heterozygous mutations in both gene locations determine the severity of abnormality seen in GT. Mutations can stop subunit production, prevent complex formation, and/or inhibit intracellular trade. When complex build-up is prevented, the subunits of αIIb or β3 are now broken. Now based on the expression and functionality of the subunits, GT is classified into three types: <5% of αIIbβ3 now specifies type I GT; now 5–20% of αIIbβ3 is type II GT; and rarely >20% of residual αIIbβ3 with dysfunctional features make up the variant type GT. Acquired GT is usually the result of autoantibody attack on platelet αIIbβ3 or isoanorbons that inhibit proper function. The production of autoantibodies has been associated with multiple hematological conditions, including immune thrombocytopenic purpura, non-Hodgkin lymphoma, multiple myeloma, myelodysplastic syndrome, hairy cell leukemia, and acute lymphoblastic leukemia, as well as platelet transfusions [21, 24, 26].

2.7 Clinical manifestations

The most common symptoms of bleeding are purpura, nosebleeds (60–80%), gingival bleeding (20–60%), and menorrhagia (60–90%). Gastrointestinal bleeding in the form of melena or hematochezia is found in 10–20% and intracranial hemorrhage is developed in 1–2%. Mucocutaneous bleeding may occur spontaneously or following minimal trauma. Epistaxis is the most common cause of severe bleeding especially in children. Menorrhagia is quite common in affected women, and there is a higher risk of serious bleeding during menarche due to the prolonged estrogenic effect on the proliferative endometrium that occurs during anovulatory cycles. Bleeding complications during pregnancy are rare; however, there is a high risk of obstetric bleeding during and after birth. Hematuria and spontaneous hemarthrosis have been described in some cases, but are generally not part of the bleeding phenotype. Currently, specific cuts could not be identified to define a positive bleeding score. Although the types of bleeding are consistent among individuals, the degree of bleeding is quite variable. The severity of bleeding (except for menorrhagia and pregnancy-related bleeding) decreases with age [8, 20].

2.8 Diagnosis

The diagnosis of GT is often not noticed, because many platelet disorders share common clinical and laboratory features. GT should be remembered in the differential diagnosis of medical history (insidious or bleeding episodes or severe bleeding after minor trauma), family history (consanguinity). In order to diagnose GT, it is necessary to choose the appropriate laboratory tests. A normal platelet count on a on a routine blood smear does not exclude the diagnosis of GT. Because patients with GT usually do not show any abnormalities in the number of platelets, complete blood count may be normal or show iron deficiency. Prothrombin time and activated partial thromboplastin time may be normal if bleeding time is prolonged; further investigations should be done [24].

Let us examine the laboratory methods in detail.

2.8.1 Complete blood count

In the evaluation of peripheral blood smear with light microscopy, normal platelet count and normal granular size should be. If the bleeding is severe and/or chronic, patients may have a red cell distribution width that increases with low hemoglobin, microcytosis, and secondary iron deficiency. Other abnormalities of the complete blood count (CBC) suggest an alternative diagnosis [20].

2.8.2 Coagulation screening tests

Prothrombin time (PT), activated thromboplastin time (aPTT), and fibrinogen values are usually normal unless a patient is evaluated in a significant acute bleeding environment and there is no evidence of consumption coagulopathy [20].

2.8.3 Platelet function screening tests

Platelet function analyzer PFA-100 provides a measure of platelet function under reduced platelets. Very long closing times (>300 s) show GT but are heat-specific. Some other disorders such as severe von Willebrand disease, Bernard Soulier syndrome, and afibrinogenemia may produce the same result. A normal PFA-100 reveals a very high negative predictive value for GT and generally excludes this diagnosis [27].

2.8.4 Platelet light transmission aggregometry

Light transmission aggregometry (LTA) is widely accepted as the gold standard diagnostic tool for evaluating platelet function. Although this test provides specific data, LTA is very time-consuming and dependent on staff and requires the use of experienced laboratories [24].

2.8.5 Whole blood impedance aggregometry

Although it can be performed using whole blood samples and lower volumes, there is insufficient evidence to support equivalent sensitivity and reproducibility compared to LTA [28].

The best way to fully diagnose GT is by mutation analysis. The genomic DNA sequence of 45 exons containing the αIIb and 3 unit should be investigated together with the junctions of the ITGB3 and ITGA2B gene and established mutations should be confirmed by a second DNA sample analysis. Genetic analysis is clinically useful for confirming the diagnosis, identifying carriers at risk, reproductive risk counseling for a particular couple/family, and definitive prenatal or preimplantation genetic diagnosis. Consequently, the diagnosis of GT involves the presence of normal platelet count (typically at the lower end of normal), long bleeding time, and long PFA time [24].

2.9 Differential diagnosis

Leukocyte adhesion deficiency type III, RASGRP2-related platelet dysfunction, BSS, Hermansky-Pudlak syndrome, von Willebrand disease, Medich platelet syndrome, Scott syndrome, and Acquired Glanzmann thrombasthenia are among the differential diagnoses [20].

2.10 Treatment

A gradual treatment standard is applied in GT treatment. The first treatment for mild bleeding is local measures including local compression, cauterization, stitching, or ice therapy. The treatment applied in case of unresponsiveness to these treatments or in heavier bleeding is antifibrinolytic therapy first, followed by platelet transfusion, and recombinant active factor VII (rFVIIa) if bleeding persists. Platelet concentrates may be single-donor and HLA-matched due to the risk of developing alloantibodies against the platelet glycoproteins, αIIbβ3, or αIIbβ3, and/or the HLA antigens. Platelet concentrates may be repeatedly transfused. If HLA-matched platelets are not found, patients should be given leukocyte-reduced platelets. This has been shown to reduce the rate of HLA immunization. Patients with severe bleeding cases should continue to receive platelet transfusion for 48 h until bleeding ceases and wound healing occurs in operated cases. These patients should be trained to avoid over-the-counter drugs that increase the risk of bleeding, such as nonsteroidal anti-inflammatories and aspirin products. Prescription drugs that may affect hemostasis should be carefully monitored [9, 24, 25, 29, 30].

Let us examine the treatment of GT according to the frequently observed conditions.

2.10.1 Treatment of minor to moderate bleeding

Local measures and/or antifibrinolytic drugs can stop mild to moderate bleeding. Local measures include compression, gelatin sponges, fibrin sealants, and topical thrombin. Antifibrinolytic agents include epsilon aminocaproic acid and tranexamic acid. Since both agents can be given orally or intravenously, they have been used successfully in the treatment of nosebleeds, bleeding gums, and menorrhagia, as well as prophylaxis before tooth extraction and other minor surgical procedures. Antifibrinolytic agents, such as tranexamic acid, can be used as a mouthwash for gingival bleeding. Antifibrinolytic use in cases of hematuria should be avoided due to the risk of a clot in the urinary tract and should be used with caution in patients undergoing procedures at high risk of thrombosis [26].

2.10.2 Treatment in epistaxis

One of the most common bleeding symptoms in GT patients is epistaxis. Local compression to epistaxis, application of tampons to the nose, topical thrombin, antifibrinolytics, and a combination of these may respond. If bleeding persists, further treatments with platelets transfusion and/or rFVIIa should be given. Antifibrinolytic agents, nasal cautery, rFVIIa, and nasal packing with synthetic materials may be used to control bleeding. If these treatments fail nasal packing with salt pork strips may be successfully used for life-threatening nasal hemorrhage in a child with GT [6, 31].

2.10.3 Treatment of menorrhagia

Antifibrinolytic agents should be first-line therapy to control menorrhagia. If it fails, hormone supplementation either progesterone alone or progesterone with estrogen may be given. A continuous estrogen-progestin oral contraceptive agent or intramuscular depot medroxyprogesterone acetate regimen given every 3 months in women with GT has been used successfully. It can be tried on hormonal intrauterine devices to reduce bleeding. Severe menorrhagia, which may be seen in many women with GT, can be treated with high-dose conjugated estrogen intravenously for 24–48 h and later by following with a combination of high doses of oral estrogen-progestin. Intensive menstrual bleeding does not always respond to typical treatment. rFVIIa has been utilized with anecdotal success in GT when anti-fibrinolytics and platelet transfusions did not control excessive menorrhagia. In addition, surgical treatments such as hysterectomy or endometrial ablation in treatment-resistant severe menorrhagia are therapeutic options [24].

2.10.4 Treatment of postpartum hemorrhage

Pregnant women with GT have high complications and are best managed in a specialized center with a multidisciplinary team. Although most complications are associated with bleeding and occur during delivery, treatment of pregnant GT patients should start in the prenatal period. According to the recommendations in the guidelines, platelet transfusions, or rFVIIa in combination with an antifibrinolytic can be used as a prophylaxis for vaginal delivery. A systematic review of 35 pregnant women with GT showed that hemorrhage during or after delivery is common and severe, and occurred up to 20 days postpartum. If patients were not given any platelet transfusions as prophylaxis, they were more likely to experience postpartum hemorrhage (63% versus 38%). The use of rFVIIa as prophylaxis was documented in three pregnancies, and it did not prevent hemorrhage in those cases. A study showed that maternal platelet alloantibodies were documented in 16 pregnancies, and plasma exchange successfully reduced the alloantibody titer in one case. Four of the 16 cases resulted in neonatal deaths, 3 of which resulted from intracranial hemorrhage between 24- and 31-weeks' gestation. One study reported successful use of rFVIIa for permanent postpartum hemorrhage. In one study, the patient was followed up with the diagnosis of GT and 18 units of random platelet concentrates, 6 units of apheresis platelet concentrates, and 2 units of erythrocyte suspension were given in the peripartum period. Although various forms of treatment have been reported about the treatment of obstetric bleeding occurring during and after birth of women pregnant with GT, there is no consensus on the most appropriate treatment. Further studies on this subject are needed [21, 26, 32].

2.10.5 Role of transfusions in the therapy of GT

Platelet transfusion allows partial correction of functional defect in patients with GT. Platelet transfusion is the standard prophylaxis when local precautions and/or antifibrinolytics cannot control bleeding and the patient is undergoing major surgery. It is not uncommon for patients with severe bleeding after trauma or delivery to require multiple platelet transfusions. Multiple platelet transfusions can be performed if necessary. An important risk associated with platelet transfusion is the possibility of developing isoantibodies. Up to 30% of patients develop anti-GPIIb/IIIa or anti-HLA antibodies after platelet transfusion. Platelet alloimmunization can lead to relative or absolute platelet refractory, causing rapid destruction of platelets and therapeutic failure of platelet transfusions. For this reason, platelet transfusions should be reserved for only major surgeries, life-threatening bleeding, and significant bleeding that does not respond to the above measure. When possible, platelet concentrates should be single-donor derived and HLA-matched. If HLA-matched platelets are not available, patients should be given leukocyte-reduced platelets because this has been shown to reduce the rate of HLA immunization. Transfusions in women of reproductive age should ideally be avoided as the antibodies can cross the placenta and affect the fetus [13, 21, 22, 33–35].

2.10.6 Use of rFVIIa in GT

Treatment of rFVIIa in a GT patient was successfully used for severe and uncontrolled bleeding in a 2-year-old child in 1996 for the first time. The worldwide use of rFVIIa continued afterward, and it was observed that most patients with GT were effective in successfully controlling bleeding. But it was also observed that it was not effective in all GT patients. The mechanism of rFVIIa is not fully delineated. It is thought are poorly attached to the surface of platelets and increase the activation of factor IX and X, thereby increasing thrombin production. Increased amount of thrombin increases platelet adhesion and supports platelet aggregation, including those not containing GPIIb/IIIa [6, 25, 33].

High success rates and relatively low risks associated with the use of rFVIIa as a treatment or prevention of bleeding in GT patients have yielded good results, especially in those who are refractory to platelet transfusion or have antiplatelet antibodies. HLA-compatible platelets have been used in the past and have been recommended as prophylaxis for major surgical procedures, including cesarean section. rFVIIa can be used to completely prevent platelet transfusion, which will reduce the risk of platelet alloimmunization in case of life-threatening bleeding when local measures and antifibrinolytics fail. The optimal dosage for use in GT patients has not been established. However, the recommended dose is bolus injections of 90 mcg/kg intravenously 3 times a day or every 2 h until bleeding stops, followed by one or more maintenance doses [6, 8, 21, 24, 33, 36, 37].

The adverse or thromboembolic events have not been reported in patients given the rFVIIa bolus. The incidence of thrombotic events is not known in GT patients treated with rFVIIa. Controlled clinical trials are needed to further assess risk [26, 36].

A UK study showed that rFVIIa was successful in 71% of patients treated within 12 h of onset, but only after 12 h, only 18% of patients responded to rFVIIa. Therefore, rFVIIa should be administered as early as possible in bleeding episodes. Minor surgeries in GT patients have been successfully treated by rFVIIa prophylaxis without the need for platelet transfusion. rFVIIa prophylaxis used is recommended by the United Kingdom Hemophilia Centre Doctors' Organization for minor surgical prophylaxis including dental extractions [6, 26, 33].

2.10.7 Other treatments

Desmopressin (DDAVP) causes VWF, FVIII, and tissue plasminogen activator to be released into the plasma. Although DDAVP is successful in treating other platelet disorders, there is little data to support its use in GT patients [26].

Rituximab (anti-CD20) is a human-mouse chimeric monoclonal antibody that targets the B cell CD20 antigen. Successful treatment has been reported for acquired GT patients. Multiple case reports have demonstrated the efficacy of rituximab in patients with treatment-resistant GT and bleeding symptoms or ecchymosis [38].

Bevacizumab (Avastin) is an anti-VEGF antibody used in combination with chemotherapy in various cancers. A single case report in the literature documented success using bevacizumab in a patient with type I GT who had severe, recurrent GI bleeding due to angiodysplasia. The patient was resistant to platelet transfusion, tranexamic acid, and embolization, but responded to bevacizumab [25].

Hematopoietic stem cell transplantation (HSCT) provides a treatment for patients with severe, recurrent bleeding episodes and resistant cases to platelet transfusion due to platelet alloantibodies. There is currently no clearly defined algorithm for transplantation in GT, and HSCT is rarely used for GT. The first successful bone marrow transplantion in GT was performed in a 4-year-old child with anti-GPIIb/IIIa antibodies in 1985. It has been reported in the literature that successful stem cell transplantation has been performed in 19 severe GT patients [26, 33, 39].

2.10.8 Future therapy

Gene therapy is very promising for GT patients to provide a treatment with significant progress using different techniques, vectors, and model organisms [40–42].

3. Bernard-Soulier syndrome

3.1 Definition

BSS is a rare autosomal recessive platelet dysfunction that is characterized by a low levels, absence, or dysfunction of the GpIb/V/IX complex on the platelet surface. BSS thrombocytopenia <20,000/mm^3 is characterized by decreased platelet adhesion, abnormal prothrombin consumption, and low-surviving large platelets. Mucocutaneous hemorrhages such as purpura, epistaxis, oral mucosa bleeding, GIS bleeding, and menorrhagia are generally seen in BSS as in other platelet function disorders [43, 44].

3.2 History

BSS with autosomal recessive transition was first described by Bernard and Soulier in 1948 as congenital bleeding disorder characterized by thrombocytopenia and large platelets [45].

3.3 Etiology

Mutations in GP1BA [GPIbα], GP1BB (GPIbß), and GP9 (GPIX) cause BSS. There of the four genes encode for the subunita of the GP Ib-IX-V complex. This key platelet receptor constituted of four subunits, GPIbα, GPIbß, GPIX, and GP5 (GPV), which included in the ratio 2:4:2:1 in endoplasmic reticulum. They because mature in Golgi apparatus before localizing at the cell surface. The GPIb-IX-V complex can attach to von Willebrand factor, fitting together like a lock and its key. Von Willebrand factor is located on the inside surface of blood vessels when there is an injury. These platelets form clots, plugging holes in the blood vessels to help stop bleeding. Due to the specified conditions occurring in BSS, clot formation is impaired and excessive bleeding occurs [46–50].

3.4 Epidemiology

The incidence of BSS is estimated to be 1 in 1 million live births, but is likely to be higher since it is often misdiagnosed [50]. In a study with 97 BSS patients in Iran, consanguineous marriage was reported in 81% of the cases' families [51].

3.5 Genetic basis

BSS occurs as a result of homozygous or compound heterozygous mutations that affect the expression of genes encoding GPIbα, GPIbß, and GPIX proteins. Two types of mutations have been reported in the GP Ib-IX-V complex. The first one is biallelic mutations, often homozygous mutations. It is characterized by a severe decrease or absence of the GP Ib-IX-V complex. To date, more than 50 biallelic mutations have been identified in the GPIbα, GPIbß, and GP9 gene. In a few cases, there is a compound heterozygous mutation. Most of the mutations identified are missense and nonsense mutations. Most BSS mutations occur in the GPIbα gene, and most of these mutations lead to a decrease in GPIbα expression on the platelet surface, and some to a loss of function. GPIbα is connected to GPIbß by disulfide bond. These are connected by noncovalent bonds with GPIX and GPV. GPV is the proteolytic subunit in this complex, and its extracellular part is destroyed by GPIbα-bound thrombin activates platelets. As a result, mutations in GP1BA, GPIbß, and GP9 in humans generally lead to a decrease in the total expression of the GP Ib-IX-V complex on the platelet surface and the disease occurs [6, 43, 44, 50, 52].

3.6 Pathophysiology

Platelets play a critical role in normal primary hemostasis and clot formation. There are specific GP receptors on the platelet membrane, which function in platelet adhesion, activation, and aggregation. The GPIb-IX-V receptor complex is responsible for platelet adhesion through its interaction with von Willebrand factor on the exposed subendothelium. The GPIb-IX-V receptor complex is composed of four transmembrane polypeptide subunits-disulfide-linked alpha and beta subunits of GPIb, and noncovalently bound subunits GPIX and GPV. The platelets of BSS cases lack or have a dysfunctional GPIb-IX-V receptor. This results in defective adhesion to the subendothelium. The dysfunctional platelets found in BSS can result from one of several different glycoprotein mutations such as missense, nonsense, or deletion mutations of the GPIbα, GPIbβ, or GPIX genes [53].

3.7 Clinical manifestations

As with other inherited platelet disorders, BSS can manifest with a tendency to bleed in early childhood. Mucocutaneous bleeding is seen predominantly. Easy bruising, purpura, epistaxis, bleeding gums, menorrhagia, and excessive bleeding after surgery or trauma are common symptoms. Menorrhagia is an important problem for female BSS patients. Prolonged menstruation may be the first symptom to help diagnose BSS in some patients. Although the severity of bleeding is associated with a genetic defect that affects receptor function and platelet count, it is highly variable in patients with the same mutations. Although bleeding sites are well defined for BSS, it is difficult to predict the severity of bleeding in patients with BSS. In some cases, no serious bleeding is observed and diagnosis may not be established until adulthood. Other genetic differences and acquired conditions affecting hemostasis are thought to affect the severity of bleeding in these patients, studies related to this need to be done. Heterozygotes may not have signs of bleeding, but giant platelets may appear in peripheral blood smear [6, 43, 46, 50, 54, 55].

3.8 Diagnosis

Although thrombocytopenia is generally observed in BSS, the number of platelets is variable. The platelet count typically ranges from 30 to 200 × 10^3/μL. Giant platelets are seen in peripheral blood smear (**Figure 1**). In order to the differential diagnosis of other giant platelet syndromes, leukocyte counts and morphology should be carefully examined. Skin bleeding time and PFA-100 closure time are found to be prolonged. Routine coagulation tests should be found normal. Prothrombin consumption and thrombin generation tests are found markedly decreased because of the defective binding of FXI and thrombin. Results of platelet aggregation studies are pathognomonic for BSS. In vitro platelet aggregation studies characteristically indicate that aggregation with ristocetin failed and responded slowly with low doses of thrombin. Flow cytometric analysis of platelet also show characteristic for BSS normal binding with CD41 (GPIIb) and CD61 (GPIIIa) antibodies, but defective binding with CD42a (GPIX), CD42b (GP Ib), CD42c (GP Ib), and CD42d (GPV) antibodies suggest BSS. Immunoblotting after separating components of the GP Ib-IX-V complex with sodium dodecyl sulfate-polyacrylamide gel electrophoresis (SDS-PAGE) may describe the defective fragments but needs specialized interpretation. Also, in recent years, most families are offered molecular genetic testing to identify which gene carries the mutations [6, 53, 56–59].

3.9 Differential diagnosis

GT, idiopathic thrombocytopenic purpura (ITP), von Willebrand disease, May-Hegglin anomaly, and other inherited giant platelet disorders, for example, gray platelet syndrome are among the differential diagnoses [52, 53].

3.10 Treatment

BSS treatment is generally supportive. Platelet transfusion is used to treat when surgery is needed or when there is a risk of life-threatening bleeding. The patient may develop antiplatelet antibodies due to the presence of glycoproteins Ib/IX/V,

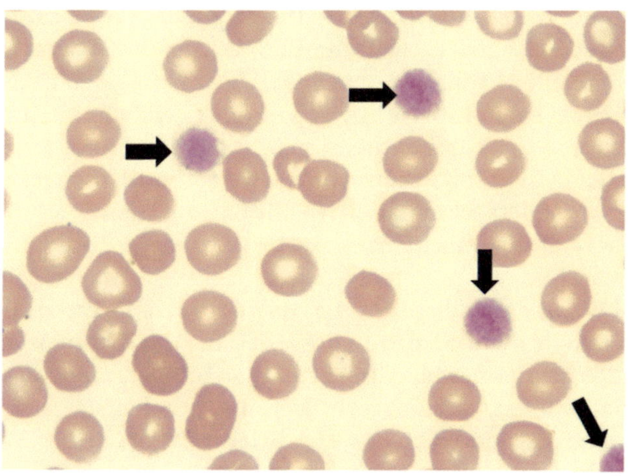

Figure 1.
Giant platelet appearance in peripheral blood smear in Bernard-Soulier syndrome.

which are present on the transfused platelets but absent from the patient's own platelets. Although some publications have suggested that patients should receive platelets from human leukocyte antigen-matched donors in order to avoid alloimmunization, this is not currently a widely accepted strategy. Antifibrinolytic agents such as p-aminocaproic acid or tranexamic acid may be useful for mucosal bleeding. rFVIIa has been reported to reduce bleeding times in patients with BSS. Desmopressin has been found to shorten bleeding episodes for some patients. A test dose should be used to determine those patients who will benefit. Stem cell transplantation has been successfully used to treat two children with BSS who had severe, life-threatening bleeding episodes. Transplantation should be considered in severe disorders when the patients have developed antiplatelet antibodies. Patients with BSS should be counseled about the importance of preventing even minor trauma as well as avoiding aspirin-containing medications and other platelet antagonists [52, 53, 60].

4. General recommendations for GT, BSS, and other inherited diseases

1. Should pay attention to dental health by brushing your teeth regularly.

2. Avoid sports activities with potential trauma (wrestling, boxing etc.).

3. Should not use salicylate and nonsteroid drugs that affect platelet function.

4. Oral contraceptives should be considered in patients with hypermenorrhoea.

5. It should be vaccinated against hepatitis A and B since blood products may be required.

6. The patient should carry a small information card describing the condition, blood group, and what to do in an emergency.

5. Conclusion

Genetic defects of the blood platelet membrane glycoproteins, GPIIb-IIIa (CD41/CD61) and GPIb-IX-V (CD42) are the origin of several rare bleeding disorders, the best known of which are GT and BSS. GT results in defective or absence of GPIIbIIIa. As a result of this, patients with GT are unable to undergo platelet aggregation, a critical step in stemming blood flow. Either gene can be affected and mutations leading to lack of expression or to expression of poorly functional forms have been identified. BSS occurs due to defective or absence of GPIb-IX-V. As a result of this, platelets from patients with BSS are unable to adhere to the damaged vessel wall at high-shear stress and also have a reduced platelet response to thrombin.

Since GT and BSS are rare diseases, diagnosis of patients can be delayed. When diagnosed early, patients will be able to prevent bleeding that may occur due to protective measures. If there is bleeding after easy bruising, mucous and oral cavities, menorrhagia, tooth extraction, tonsillectomy, or other surgical interventions, GT or BSS should be considered among the differential diagnoses. Although GT cannot be diagnosed with routine laboratory tests, BSS is suspected in the presence of thrombocytopenia and giant platelet. Detailed examination is required for a definitive diagnosis. Treatment includes local measures, platelet infusion, rFVIIa, and other treatments. Although there is no permanent treatment for now, research is still ongoing. For this, it is more important for patients to avoid situations that may increase their tendency to bleed.

Acknowledgments

The author would like to thank Prof. Dr. Kenan KOCABAY, who has contributed greatly to his specialty education in medicine and has helped write this book chapter.

Author details

Muhammet Mesut Nezir Engin
Department of Child Health and Diseases, Faculty of Medicine, Düzce University, Düzce, Turkey

*Address all correspondence to: doktormesut@hotmail.com

References

[1] Kaptan K. Functional platelet disorders. Turkiye Klinikleri Journal of Internal Medical Sciences. 2005;**1**(2):31-40

[2] Fışgın T, Koca D. Inherited platelet disorders. Turkish Children's Journal of Hematology. 2011;**5**(1):1-10

[3] Neunert CE, Journeycake JM. Congenital platelet disorders. Hematology/Oncology Clinics of North America. 2007;**21**(4):66384

[4] Streif W, Knöfler R, Eberl W. Inherited disorders of platelet function in pediatric clinical practice: A diagnostic challenge. Klinische Pädiatrie. 2010;**222**:203-208

[5] Cattaneo M. Inherited platelet-based bleeding disorders. Journal of Thrombosis and Haemostasis. 2003;**1**:1628-1636

[6] Diz-Kucukkaya R. Inherited platelet disorders including Glanzmann thrombasthenia and Bernard-Soulier syndrome. Hematology. American Society of Hematology. Education Program. 2013;**2013**:268-275

[7] Poon MC, d'Oiron R. Alloimmunization in congenital deficiencies of platelet surface glycoproteins: Focus on Glanzmann's thrombasthenia and Bernard–Soulier's syndrome. Seminars in Thrombosis and Hemostasis. 2018;**44**(06):604-614

[8] George JN, Caen JP, Nurden AT. Glanzmann's thrombasthenia: The spectrum of clinical disease. Blood. 1990;**75**(07):1383-1395

[9] Krause KA, Graham BC. Glanzmann Thrombasthenia. StatPearls Publishing. 2020. Available from: https://www.ncbi.nlm.nih.gov/books/NBK538270/

[10] Nurden AT. Glanzmann thrombasthenia. Orphanet Journal of Rare Diseases. 2016;**1**:10. DOI: 10.1186/1750-1172-1-10

[11] Doherty D, Singleton E, Byrne M, Ryan K, O'Connell NM, O'Donnell JS, et al. Missed at first Glanz: Glanzmann thrombasthenia initially misdiagnosed as Von Willebrand disease. Transfusion and Apheresis Science. 2019;**58**(1):58-60

[12] Iqbal I, Farhan S, Ahmed N. Glanzmann thrombasthenia: A clinicopathological profile. Journal of the College of Physicians and Surgeons–Pakistan. 2016;**26**(8):647-650

[13] Poon MC, Di Minno G, d'Oiron R, Zotz R. New insights into the treatment of Glanzmann thrombasthenia. Transfusion Medicine Reviews. 2016;**30**(2):92-99

[14] Glanzmann E. Hereditare hamorrhagische thrombasthenie. Ein Beitrag zur Pathologie der Blutplattchen. J Kinderkranken. 1918;**88**:113

[15] Braunsteiner H, Pakesch F. Thrombocytoasthenia and thrombocytopathia. Old names and new diseases. Blood. 1956;**11**:965-976

[16] Caen JP, Castaldi PA, Lecrec JC, Inceman S, Larrieu MJ, Probst M, et al. Glanzmann's thrombasthenia. I. Congenital bleeding disorders with long bleeding time and normal platelet count. American Journal of Medicine. 1966;**44**:4

[17] Nurden AT, Freson K, Seligsohn U. Inherited platelet disorders. Haemophilia. 2012;**18**(Suppl 4):154-160

[18] Vogel ER, VanOosten SK, Holman MA, et al. Cigarette smoke enhances proliferation and extracellular matrix deposition by human fetal airway smooth muscle. American Journal of Physiology. Lung Cellular and Molecular Physiology. 2014;**307**(12):978-986

[19] Humphreys I, Saraiya S, Belenky W, Dworkin J. Nasal packing with strips of cured pork as treatment for uncontrollable epistaxis in a patient with Glanzmann thrombasthenia. The Annals of Otology, Rhinology, and Laryngology. 2011;**120**(11):732-736

[20] Botero JP, Lee K, Branchford BR, Bray PF, Freson K, Lambert MP. Glanzmann thrombasthenia: Genetic basis and clinical correlates. Haematologica. 2020;**105**(4):888-894

[21] Nurden AT, Fiore M, Nurden P, Pillois X. Glanzmann thrombasthenia: A review of ITGA2B and ITGB3 defects with emphasis on variants, phenotypic variability, and mouse models. Blood. 2011;**118**(23):59966005

[22] Nurden AT, Pillois X. ITGA2B and ITGB3 gene mutations associated with Glanzmann thrombasthenia. Platelets. 2018;**29**(1):98101

[23] Richards S, Aziz N, Bale S, et al. Standards and guidelines for the interpretation of sequence variants: A joint consensus recommendation of the American College of Medical Genetics and Genomics and the Association for Molecular Pathology. Genetics in Medicine. 2015;**17**(5):405-424

[24] Solh T, Botsford A, Solh M. Glanzmann's thrombasthenia: Pathogenesis, diagnosis, and current and emerging treatment options. Journal of Blood Medicine. 2015;**6**:219-227

[25] Stevens RF, Meyer S. Fanconi and Glanzmann: The men and their works. British Journal of Haematology. 2002;**119**(4):901-904

[26] Fiore M, Nurden AT, Nurden P, Seligsohn U. Clinical utility gene card for: Glanzmann thrombasthenia. European Journal of Human Genetics. 2012;**20**(10):1102

[27] Harrison P. The role of PFA-100 testing in the investigation and management of haemostatic defects in children and adults. British Journal of Haematology. 2005;**130**(1):3-10

[28] Al Ghaithi R, Drake S, Watson SP, Morgan NV, Harrison P. Comparison of multiple electrode aggregometry with lumi-aggregometry for the diagnosis of patients with mild bleeding disorders. Journal of Thrombosis and Haemostasis. 2017;**15**(10):2045-2052

[29] Grainger JD, Thachil J, Will AM. How we treat the platelet glycoprotein defects; Glanzmann thrombasthenia and Bernard Soulier syndrome in children and adults. British Journal of Haematology. 2018;**182**(5):621-632

[30] Temizkan RC, Engin MMN, Türen B, Kocabay K. Evaluation of pediatric patients with gastrointestinal bleeding following a single dose ingestion of NSAID. Archives of Emergency Medicine and Intensive Care. 2020;**3**(1):01-05

[31] Blickstein D, Dardik R, Rosenthal E, et al. Acquired thrombasthenia due to inhibitory effect of glycoprotein IIbIIIa autoantibodies. The Israel Medical Association Journal. 2014;**16**(5):307-310

[32] Ateş S, Molla T, Özkal F, Çetin G, Başaranoğlu G, Dane B. Peripartum management of a pregnant with Glanzmann's thrombasthenia: A case report. The Medical Bulletin of Haseki. 2014;**52**:208-211

[33] Bennett JS. Structure and function of the platelet integrin alphaIIbbeta3. The Journal of Clinical Investigation. 2005;**115**(12):3363-3369

[34] Fiore M, Firah N, Pillois X, Nurden P, Heilig R, Nurden AT. Natural history of platelet antibody formation against alphaIIbbeta3 in a French cohort of Glanzmann thrombasthenia patients. Haemophilia. 2012;**18**(3):e201-e209

[35] Siddiq S, Clark A, Mumford A. A systematic review of the management

and outcomes of pregnancy in Glanzmann thrombasthenia. Haemophilia. 2011;**17**(5):e858-e869

[36] Nurden AT, Ruan J, Pasquet KM, et al. A novel 196Leu to Pro substitution in the beta3 subunit of the alphaIIbbeta3 integrin in a patient with a variant form of Glanzmann thrombasthenia. Platelets. 2002;**13**(2):101-111

[37] Cattaneo M, Cerletti C, Harrison P, et al. Recommendations for the standardization of light transmission aggregometry: A consensus of the working party from the platelet physiology subcommittee of SSC/ISTH. Journal of Thrombosis and Haemostasis. 2013;**11**:1183-1189

[38] Morel-Kopp MC, Melchior C, Chen P, et al. A naturally occurring point mutation in the beta3 integrin MIDAS-like domain affects differently alphavbeta3 and alphaIIIbbeta3 receptor function. Thrombosis and Haemostasis. 2001;**86**(6):1425-1434

[39] Nurden AT, Pillois X, Wilcox DA. Glanzmann thrombasthenia: State of the art and future directions. Seminars in Thrombosis and Hemostasis. 2013;**39**(6):642-655

[40] Wilcox DA, Olsen JC, Ishizawa L, et al. Megakaryocyte-targeted synthesis of the integrin beta(3)-subunit results in the phenotypic correction of Glanzmann thrombasthenia. Blood. 2000;**95**(12):3645-3651

[41] Fang J, Hodivala-Dilke K, Johnson BD, et al. Therapeutic expression of the platelet-specific integrin, alphaIIbbeta3, in a murine model for Glanzmann thrombasthenia. Blood. 2005;**106**(8):2671-2679

[42] Sullivan SK, Mills JA, Koukouritaki SB, et al. High-level transgene expression in induced pluripotent stem cell-derived megakaryocytes: Correction of Glanzmann thrombasthenia. Blood. 2014;**123**(5):753-757

[43] Lambert MP, Poncz M. Inherited platelet disorders. In: Orkin SH, Fisher DE, Ginsburg D, Look AT, Lux SE, Nathan DG, editors. Nathan and Oski's Hematology and Oncology of Infancy and Childhood. 8th ed. Philadelphia: Elsevier Saunders; 2015. pp. 1167-1203

[44] Tokgöz H, Çalışkan Ü. Clinical and genotypic findings in patients with Bernard Soulier syndrome: A single center experience. Turkish Journal of Pediatric Disease. 2017;**1**:51-55

[45] Bernard J, Soulier JP. Sur une nouvelle variete de dystrophie thrombocytaire-hemorragipare congenitale. La Semaine des Hôpitaux de Paris. 1948;**24**:3217-3222

[46] López JA, Andrews RK, Afshar-Kharghan V, Berndt MC. Bernard-Soulier syndrome. Blood. 1998;**91**:4397-4418

[47] Luo SZ, Mo X, Afshar-Kharghan V, Srinivasan S, López JA, Li R. Glycoprotein Ibalpha forms disulfide bonds with 2 glycoprotein Ibbeta subunits in the resting platelet. Blood. 2007;**109**(603):609

[48] Ulsemer U, Strassel C, Baas M-J, Salamero J, Chasserot-Golaz S, Cazenave J-P, et al. Biosynthesis and post-translational processing of normal and mutant platelet glycoprotein GPIb-IX. The Biochemical Journal. 2001;**358**:295-303

[49] Li R, Emsley J. The organizing principle of the platelet glycoprotein Ib-IX-V complex. Journal of Thrombosis and Haemostasis. 2013;**11**:605-614

[50] Savoia A, Kunishima S, De Rocco D, Zieger B, Rand ML, Pujol-Moix N, et al. Spectrum of the mutations in Bernard-Soulier syndrome. Human Mutation. 2014;**35**(9):1033-1045

[51] Toogeh G, Keyhani M, Sharifian R, et al. A study of Bernard-Soulier syndrome in Tehran, Iran. Archives of Iranian Medicine. 2010;**13**(6):549-551

[52] Balduini CL, Savoia A. Genetics of familial forms of thrombocytopenia. Human Genetics. 2012;**131**:1821-1832

[53] Pham A, Wang J. Bernard-Soulier syndrome: An inherited platelet disorder. Archives of Pathology & Laboratory Medicine. 2007;**131**(12):1834-1836

[54] Andrews RK, Berdt MC. Bernard-Soulier syndrome: An update. Seminars in Thrombosis and Hemostasis. 2013;**39**(6):656-662

[55] Farhan S, Iqbal I, Ahmed N. Bernard Soulier syndrome: 10 years' experience at a tertiary care hospital. Pakistan Journal of Medical Sciences. 2019;**35**(3):705-708

[56] Bolton-Maggs PHB, Chalmers EA, Collins PW, et al. A review of inherited platelet disorders with guidelines for their management on behalf of the UKHCDO. British Journal of Haematology. 2006;**135**(5):603-633

[57] Balduini CL, Pecci A, Noris P. Diagnosis and management of inherited thrombocytopenias. Seminars in Thrombosis and Hemostasis. 2013;**39**(2):161-171

[58] Harrison P, Mackie I, Mumford A, et al. Guidelines for the laboratory investigation of heritable disorders of platelet function. British Journal of Haematology. 2011;**155**(1):30-44

[59] Miller JL. Glycoprotein analysis for the diagnostic evaluation of platelet disorders. Seminars in Thrombosis and Hemostasis. 2009;**35**(2):224232

[60] Balduini CL, Iolascon A, Savoia A. Inherited thrombocytopenias: From genes to therapy. Hema. 2002;**87**: 860-880

Chapter 6

Thrombocytopenia in Neonates

Bernhard Resch

Abstract

Thrombocytopenia defined as platelet count below 150,000/μL is not an uncommon event at the neonatal intensive care unit (NICU). In our region we calculated a prevalence of nearly 2 of 1000 live births. Early-onset neonatal thrombocytopenia (NT) occurring within the first 72 hours of life is more common than late-onset NT. Preterm infants are affected more often than term infants and bacterial infection is the most common diagnosis associated with NT. There are a lot of maternal, perinatal, and neonatal causes associated with NT and complications include bleedings with potentially life-threatening intracranial hemorrhage. Alloimmune thrombocytopenia (NAIT) often presents with severe thrombocytopenia (<30,000/μL) in otherwise healthy newborns and needs careful evaluation regarding HPA-1a antigen status and HLA typing. Platelet transfusions are needed in severe NT and threshold platelet counts might be at ≤25,000/μL irrespective of bleeding or not. Immune mediated NT recovers within 2 weeks with a good prognosis when there happened no intracranial hemorrhage. This short review gives an overview on etiology and causes of NT and recommendations regarding platelet transfusions.

Keywords: neonatal thrombocytopenia, prematurity, platelet transfusion, alloimmune thrombocytopenia, early- and late-onset thrombocytopenia, severity, intracranial hemorrhage, bleeding, incidence, mortality

1. Introduction

Thrombocytopenia, defined as a platelet count <150,000/μL or below 150×10^9/L, occurs more often during the neonatal period than in any other populations. Thrombocytopenia implicates an increased risk of bleeding, and is associated with significant morbidity mainly due to intracranial hemorrhage. As a result, it is important to identify infants at risk, and if indicated (see below recommendations for platelet transfusion), to initiate therapy to prevent complications.

Recently our research group on neonatal infectious diseases and epidemiology at the Medical University of Graz, Austria, published retrospectively collected data on neonatal thrombocytopenia (NT) [1]. Of 371 neonates diagnosed as having NT, 312 (84.1%) had early-onset NT (EOT) and 59 (15.9%) had late-onset NT (LOT) defined as NT before or after 72 hours of age, respectively. The degree of NT was defined as mild, platelet counts of 100,000–150,000/μL; moderate, counts of 50,000–<100,000/μL; severe, counts of 30,000–<50,000/μL; and very severe, counts of <30,000/μL, according to the description of Wiedmeier et al. [2]. The majority—nearly three-fourth of the cohort—had mild (33%) to moderate (38%) NT; only 14% had severe and 15% very severe NT [1]. Seventy-six percent of the neonates were born preterm and this rate was approximately the same for either EOT or LOT (76 and 77%, respectively). The percentage of extremely low gestational age newborns (ELGAN, below 28 weeks of gestational age) was 20% in total. The incidence of NT in preterm infants was 4% (282/6964) in our population during the years 1990–2012. Thus, we calculated the prevalence of NT as being 1.8/1000 live births in our region (Southern Styria with around 200,000 live births during the study period) [1].

A total of 40 neonates (10.8%) died; 36 (90%) had EOT, and 4 (10%) had LOT; and 30 (75%) neonates were still thrombocytopenic at the time of death. Interestingly, bleeding signs were significantly associated with mortality in our study. On the other hand, severity of NT was not associated with mortality. Only cutaneous bleedings were found to be associated with severity of NT. The mean duration of NT was significantly longer in case of LOT compared to EOT (8.9 vs. 16.8 days; $p < 0.001$); and the duration was positively correlated with severity of NT. At least we found that platelet transfusion did not shorten the duration of NT [1].

2. Etiology and causes of neonatal thrombocytopenia

Thrombocytopenia is present in 1–5% of newborns at birth, and severe thrombocytopenia defined as platelet count below 50,000/μL occurs in 0.1–0.5% [3]. But thrombocytopenia is more common in neonates needing intensive care at the neonatal intensive care unit (NICU) with rates up to 50%. Every fifth newborn is at risk at the NICU to develop thrombocytopenia, and 8% of preterm and 6% of term infants are at risk for severe thrombocytopenia [4].

Main mechanisms of thrombocytopenia include increased platelet consumption and/or sequestration, and often neonatal thrombocytopenia is of multifactorial origin. Thus, on the one hand there is rapid consumption like in case of necrotizing enterocolitis (NEC) and on the other hand slow recovery by impaired platelet production.

There are a lot of maternal, perinatal, and neonatal causes that might be associated with the occurrence of NT. **Figure 1** shows the main features and causes of NT by separating early- and late-onset thrombocytopenia [5].

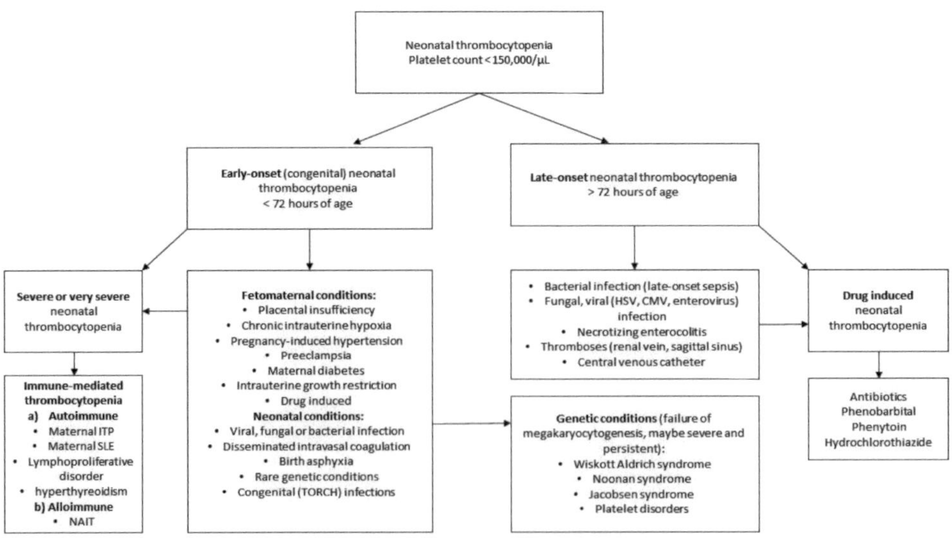

Figure 1.
Etiology and causes of early- and late-onset thrombocytopenia.

Austria [1]	2018	Tunisia []	2016	Turkey []	2013
371 cases	23-year period	112 episodes	4-year period	134 cases	5-year period
EOS	128 (34)	LOS	29 (19)	Sepsis	45 (34)
Asphyxia	95 (25)	IUGR	26 (17)	IUGR	25 (19)
LOS	47 (13)	EOS	23 (15)	Preeclampsia	13 (9.7)
NEC	16 (4.1)	DIC	22 (14)	Maternal thr.	6 (4.5)
Chrom. aberration	15 (3.9)	PIH	18 (12)	Drug	5 (3.7)
HDN	9 (2.4)	Unexplained	15 (9.6)	Hydrops fetalis	4 (3.0)
CMV	9 (2.4)	Cong. rubella	6 (3.8)	Perinatal asphyxia	4 (3.0)
MPD	6 (1.6)	Asphyxia	5 (3.2)	Ablatio placenta	4 (3.0)
NAIT	4 (1.0)	NEC	3 (1.9)	Drug + sepsis	4 (3.0)
K-M-syndrome	2 (0.5)	Tris 21	2 (1.3)	Rh incompatibility	4 (3.0)
Metab. disorders	2 (0.5)	CMV	2 (1.3)	NEC	3 (2.2)
Thrombosis	2 (0.5)	Exchange transf.	2 (1.3)	Maternal ITP	3 (2.2)
		Toxoplasmosis	1 (0.7)	Congenital anomaly	2 (1.5)
		Neonatal lupus	1 (0.7)	Tris 21	2 (1.5)
		Maternal ITP	1 (0.7)	NAIT	2 (1.5)
				HELLP syndrome	2 (1.5)
				Metab. disorders	2 (1.5)
				Maternal GDM	2 (1.5)
				Neonatal jaundice	1 (0.8)
				Mother with SLE	1 (0.8)

Data are given as number (%).
EOS: early-onset sepsis, LOS: late-onset sepsis, NEC: necrotizing enterocolitis, HDN: hemolytic disease of the newborn, CMV: cytomegalovirus, MPD: myeloproliferative disease, NAIT: neonatal alloimmune thrombocytopenia, K-M-syndrome: Kasabach-Merritt syndrome, DIC: disseminated intravasal coagulation, IUGR: intrauterine growth restriction, GDM: gestational diabetes mellitus, SLE: systemic lupus erythematodes.

Table 1.
Neonatal diagnoses associated with thrombocytopenia from three studies (Austria [1], Tunisia [6], and Turkey [7]).

Own data on neonatal diagnoses associated with thrombocytopenia retrospectively collected over 23 years compared to data from Tunisia and Turkey—in order to give a broader view on dominant causes and diagnoses—are shown in **Table 1** [1]. Main diagnoses were early- and late-onset sepsis, and intrauterine growth restriction. Other dominant features like asphyxia and NEC differed between centers as did more sporadic causes of NT.

An association of NT with bacterial infection is well known. Rates have been reported of 30% [7], 36% [8], and 47% [1]. And severe NT and very severe NT were commonly associated with sepsis [1, 8]. Late-onset NT often is reported to result in prolonged courses of NT [1, 9, 10].

Birth asphyxia is a common diagnosis associated with NT and reported up to 25% [1, 7, 8]. NEC is a morbidity often complicated by NT. Resch et al. reported a rate of 4.1% [1] and that was twice as high as reported elsewhere [7, 8]. Other associations, including chromosomal anomalies, metabolic disorders, and thromboses, ranged between 0.5 and 3.9% in the literature [1, 6–8]. Von Lindern et al. [8] reported on a 10% rate of hemolytic disease of the newborn (HDN), which was four times higher compared to the rate of 2.4% reported by Resch et al. [1].

In preterm infants, NT is rarely diagnosed at low rates between 4 and 12% [1, 10]. Most studies report higher rates of NT ranging between 22 and 35% [2, 3, 8, 11–13], and highest rates (53–70%) have been reported from developing countries [14, 15]. Own data revealed that 75% of a cohort of thrombocytopenic neonates were preterm neonates. The association between NT and prematurity or low birth weight is well documented [3, 6–8, 11, 16, 17]. In this context small-for-gestational age (SGA) is a well-known risk factor for developing NT [11, 16, 17], and rates have been reported as high as 30–53% [1, 17, 18].

3. Pathomechanisms of neonatal thrombocytopenia

3.1 Immune-mediated thrombocytopenia

One possibility of low platelet counts is increased destruction that is observed in several neonatal conditions. Placental crossing of maternal antibodies is the cause of NT in case of immune-mediated NT, which destroys neonatal platelets. Immune-mediated processes are very common causes of neonatal thrombocytopenia, and the antibodies responsible may be autoantibodies, drug-dependent antibodies, or alloantibodies. The mechanism behind is an interaction with platelet membrane antigens or the formation of immune complexes, which can bind to reticuloendothelial cell Fc receptors. As a result platelets become cleared from blood vessels [19].

3.1.1 Neonatal alloimmune thrombocytopenia (NAIT)

In NAIT, fetal platelets contain an antigen inherited from the father that the mother lacks. The mother produces antiplatelet antibodies from the immunoglobulin G (IgG)-type against the platelet antigen during pregnancy that is recognized as being foreign. Thereafter IgG antibodies cross the placenta and destroy fetal platelets that express the paternal antigen [20].

3.1.2 Neonatal autoimmune thrombocytopenia

It is mediated by maternal autoantibodies that react with both maternal and fetal platelets. This occurs in maternal autoimmune disorders, including immune thrombocytopenia purpura (ITP) and systemic lupus erythematosus (SLE) [20].

3.1.3 Drug-induced immune thrombocytopenia

Drug-induced immune thrombocytopenia is typically caused by platelet destruction from maternal drug-dependent antibodies and, rarely, by neonatal antibodies. Bone marrow suppression also can result in thrombocytopenia due to decreased platelet production. Neonatal drug-induced immune thrombocytopenia is usually caused by maternal drug-dependent antibodies formed after drug exposure to the mother during pregnancy [19, 20]. Maternal antibodies can cross the placenta and affect fetal and neonatal platelets. This mechanism is similar to that seen in mothers with primary immune thrombocytopenia purpura (ITP). Drugs associated with maternal drug-mediated platelet destruction include quinine, quinidine, trimethoprim-sulfamethoxazole, vancomycin, penicillin, rifampin, carbamazepine, phenytoin, valproic acid, ceftriaxone, ibuprofen, mirtazapine, oxaliplatin, suramin, GP IIb/IIIa inhibitors (e.g., abciximab, tirofiban, eptifibatide), and heparin. Some drugs may cause thrombocytopenia at the initial exposure without prior sensitization. This commonly occurs with the glycoprotein IIb/IIIa inhibitors and has also been seen with other drugs, such as vancomycin and piperacillin [21]. Rarely, platelet destruction can be caused by neonatal drug-dependent antibodies, such as seen in heparin-induced thrombocytopenia (HIT). HIT antibodies can promote thrombosis by inducing platelet activation. Patients with suspected HIT require immediate institution of a non-heparin anticoagulant [20].

3.1.4 Non-immune drug-induced NT

Many drugs used as chemotherapy cause thrombocytopenia by bone marrow suppression (*non-immune drug induced NT*). Antibiotics, such as linezolid, daptomycin, and valacyclovir can also cause moderate thrombocytopenia in some patients by suppression of platelet production [20].

3.2 Thrombopoiesis

Platelet production—thrombopoiesis—is initiated by a thrombopoietic stimulus, and the most important stimulant is the chemokine thrombopoietin (TPO) besides several cytokines and chemokines that are also involved in the process (e.g., IL-3, IL-6, IL-11, GM-CSF, stromal cell-derived factor-1 and fibroblast growth factor 4) [22]. TPO promotes the proliferation of megakaryocyte progenitors and the maturation of megakaryocytes. These mature megakaryocytes are at least responsible for generation and release of new platelets into the blood vessels [22].

3.2.1 The homeostasis of TPO levels

The homeostasis of TPO levels is regulated by the thrombopoietin c-Mpl (myeloproliferative leukemia protein) receptor-mediated uptake and destruction of the hormone with the aim to have steady-state amounts of hepatic TPO. When bound to the platelet c-Mpl receptors, the hormone gets removed from the circulation and blood levels are reduced. In case of inflammatory processes IL-6 is released from macrophages and fibroblasts (via TNF-α) and enhances hepatic TPO production [22].

Another phenomenon adding to the steady-state model of TPO regulation is the physiological response to severe thrombocytopenia of bone marrow stromal cells, which under normal circumstances produce low amounts of TPO-mRNA but increase transcription markedly in case of thrombocytopenia [22]. IL-6, stimulated by inflammation processes, leads to increased levels of TPO resulting in reactive thrombocytosis; and TPO is now confirmed as the final mediator of inflammation-induced thrombocytosis [22].

There are similarities and differences between neonatal and adult thrombocytopenia. Plasma TPO concentrations are known to be higher in healthy neonates compared to healthy adults. But in NT TPO levels are lower even when adult thrombocytopenia has the same degree [19]. Interestingly, megakaryocyte progenitors of neonates have a higher proliferative potential than those of adults resulting in larger megakaryocytes. Neonatal megakaryocyte progenitors are more sensitive to TPO both in vitro and in vivo than adult progenitors [19]. Cells are present both in the bone marrow and the peripheral blood of neonates in contrast to adult ones that are almost exclusively present in the bone marrow [19] (**Figure 2**).

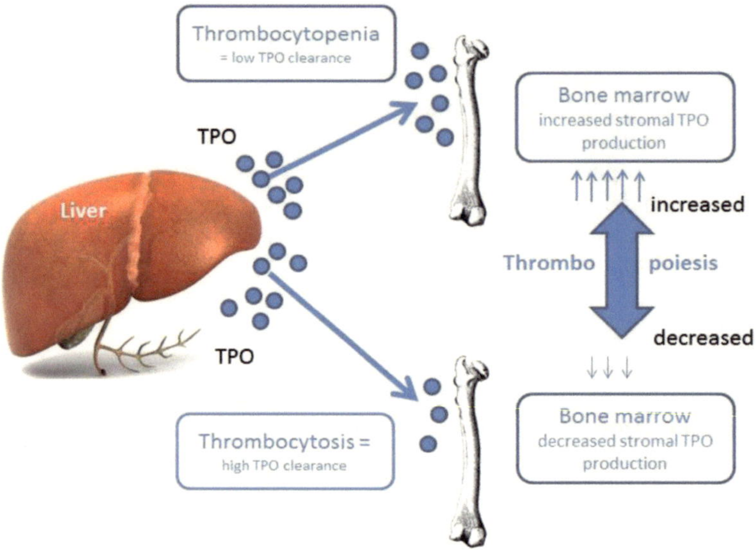

Figure 2.
A simplified model of TPO regulation.

3.2.2 Clinical conditions and their pathomechanisms

Chronic intrauterine hypoxia is commonly observed when associated with placental insufficiency due to all conditions of pregnancy-induced hypertension (hypertension alone or pre-eclampsia or hypertension-elevated liver enzymes-low platelet counts—HELLP—syndrome, and gestational or maternal diabetes) and is usually manifested by fetal intrauterine growth restriction and hematological abnormalities including NT. The pathomechanisms behind are not completely understood but lower levels of megakaryocyte progenitors have been found that increased during normalization of NT [23].

The hematopoietic microenvironment plays a significant role in chronic hypoxia-induced suppression of megakaryocytopoiesis, and, not astonishingly, preterm infants' megakaryocyte progenitors are more vulnerable to ischemic insults than progenitors from term neonates or adults [24].

Overall, observations demonstrated that thrombopoiesis is up-regulated in *neonatal sepsis and/or NEC*, but this effect can also be down-regulated resulting in "hypoproliferation" [19]. Platelet factor 4 is a potent inhibitor of megakaryocyte proliferation that is released from activated platelets during severe sepsis. This regulates neonatal megakaryocytopoiesis negatively [25].

In *HIV-associated thrombocytopenia*, evidence that splenic platelet sequestration decreased platelet production had been observed despite larger megakaryocyte mass [26]. In neonates, ineffective platelet production was described as being the main mechanism of HIV-associated thrombocytopenia [27]. The mechanisms in other *congenital infections* of the TORCH complex mostly remain to be speculative despite their common association with NT [20].

4. Neonatal alloimmune thrombocytopenia (NAIT)

Harrington et al. first described in 1953 two infants born with severe thrombocytopenia to mothers with normal platelet counts [28]. Both newborns recovered despite severe bleeding and other complications after 2 and 8 weeks, respectively.

This immunological disease now is well described as neonatal alloimmune thrombocytopenia (NAIT).

In 1962 a maternal antibody against a platelet alloantigen was detected causing NAIT [29]. This platelet alloantigen determined as PlA1 was the cause of platelet destruction in two of the newborns reported in the study by Shulman et al. [29]. Later, PlA1 was found to be identical to an antigen called Zwa [30] and now is known as human platelet antigen 1a (HPA-1a). Over the following years several other platelet-specific antigens were detected as being able to induce maternal immunization during pregnancy with subsequent fetal platelet destruction, thus, being an important complication of pregnancy with diagnostic and therapeutic challenges [31].

The incidence of NAIT calculated from large studies on women negative for HPA-1a lies in between 1 in 1000–2000 HPA-1a positive newborns [32–34]. The incidence of HPA-1a negative phenotype is about 2.5%; and one-third is at high risk to get immunized in case of a HPA-1a positive fetus, and this association is triggered by HLA-DR antigen B3*0101 positivity [31].

The main problem of NAIT is that it can lead to serious bleedings including intracranial hemorrhage and death. In full term infants it is the leading cause of intracranial hemorrhage [35]. Other clinical findings are petechiae or purpura associated with very low platelet counts without any explanation (after exclusion of bacterial and viral infection—TORCH complex, or disseminated intravascular coagulation). A previous history of NAIT results in more severe disease. Around 10–20% of the newborns have intracranial hemorrhages, and the vast majority of 80% occurs already before birth. After birth the greatest risk of bleeding is in the time span of the first 4 days of life. Untreated, NT resolves within 2–3 weeks [35].

Even in mildly affected infants serological investigation including ABO, HPA and HLA typing (further details are beyond the scope of this chapter) is indicated because results can be critical for effective management of future pregnancies. For the most informative evaluation, it is important to study blood samples from both mother and father.

A systematic review on incidence and consequences of NAIT reported on 6 of 21 studies (full text analysis) from initial 768 studies [36]. Nearly 60,000 newborns were screened, with severe thrombocytopenia in 89 cases (0.15%); and NAIT was diagnosed in 24 of these 89 newborns (27%) resulting in an incidence of 1:2500. Six newborns (25%) had diagnosis of intracranial hemorrhage and most likely of antenatal origin. Hence, intracranial hemorrhage due to NAIT occurred in 1:10–11,000 newborns [36, 37]. This is the most severe complication having a 1–7% risk of death. Survivors are known to have sequelae including mental retardation, cerebral palsy, cortical blindness and seizures in 7–26% of pregnancies [38]. In contrast to ABO- or Rh-incompatibility, immunization occurs often during the first pregnancy.

Severity of NAIT is associated with parity (second pregnancy often more severe), HPA-a1 antibodies level, outcome of a former pregnancy, and the type of alloimmunization (HPA-1a more severe than HPA-5b) and the HLA type: homozygote HPA-1b and negative for HLA DRB3*0101 leads to no NAIT with negative predictive value of 99.6%, but homozygote HPA-1b plus HLA DRB3*0101 positivity leads to NAIT with a positive predictive valuc of 35% [39] Interestingly, antenatal IVIG is again more effective depending on HLA status [31]. Also of importance is the risk of recurrent intracranial hemorrhage of being 80–90%.

4.1 How to proceed in case of suspected NAIT?

At first the medical history should be looked for a previous neonate with thrombocytopenia of unknown origin and/or having platelet count below 50×10^9/L and/or previous history of intracranial hemorrhage of uncertain origin [40]. Is the mother thrombocytopenic, one should evaluate for maternal anti-platelet auto-antibodies or a history of immune thrombocytopenia. Is the mother not thrombocytopenic, maternal and paternal platelet antigen typing and maternal platelet HPA antibody testing should be done. In case of incompatibility at HPA loci (1–6, 9, 15) and presence of specific maternal HPA-antibody, diagnosis of NAIT is given. Are there no incompatibilities and no anti-platelet antibodies or only nonspecific antibodies of the mother present, then no further evaluation is necessary besides maternal antibodies against paternal platelets are positive (preferred at 30 weeks' gestation). A third variant is a positive incompatibility without maternal anti-HPA antibodies (provided no reaction with paternal platelets), then there is no further evaluation necessary [40].

During pregnancy, strategies using IVIG and corticosteroids have been successful. The success rate with IVIG alone was reported to be as high as 98.7% and is in line with a Cochrane analysis reporting 97.3% [37, 41]

5. Complications of neonatal thrombocytopenia

5.1 Bleedings

The prevalence of hemorrhages in thrombocytopenic neonates is approximately 20–30% according to the literature [16, 42]. The risk of hemorrhage is associated with lower gestational age, definite causes of thrombocytopenia, and the severity of concomitant morbidities [8, 9, 43].

An observational study including 169 neonates with severe NT identified severe sepsis and NEC as the most common diagnoses associated with bleeding neonates [35]. In those neonates with mild or no hemorrhage, the most common cause of severe NT has been documented as being intrauterine growth restriction and maternal pregnancy-induced hypertension [44].

A causal link between thrombocytopenia and intracranial hemorrhage is not known. Interestingly, platelet transfusions could not reduce the risk of intracranial bleedings [1, 45]. The majority of preterm neonates with severe intraventricular hemorrhage (IVH) becomes thrombocytopenic during the course of bleeding, thus, thrombocytopenia might not be the cause of IVH [46, 47]. Additionally, considering IVH as a multifactorial event, it seems highly unlikely that an isolated low platelet count leads to bleeding [44].

Another point of interest is the fact that comparable rates of IVH have been reported independent of the severity of NT [17]. In contrast, cutaneous bleeding conditions have been associated with the severity of NT [1, 46], and the prevalence of skin bleeding in thrombocytopenic neonates has been reported as being as high as 81% [48].

5.2 Mortality

The association of increased mortality rates with increasing numbers of platelet tranfusions mainly reflects the severity of the underlying disease or condition, for example, extremely low gestational age newborn. These infants are known to be at high risk for severe IVH that again is associated with high risk of death. Additionally, the more severe NT is the higher is the rate of mortality, and some data suggest that NT contributes to mortality rather than simply being a measure of disease severity [5]. Three studies from the USA, the UK, and Mexico reported higher mortality rates in neonates who had received platelet transfusions compared with those who had not [49–51]. The direct effects of platelet transfusions are questionable, as specific effects have not been properly evaluated, and the influence of preexisting morbidity is difficult to evaluate [44]. But as shown by Curley et al. [52] (see below), waiting until platelets have fallen below 25 × 10^9/L before indicating transfusion is of more benefit than earlier transfusion at cut-off level below 50 × 10^9/L. Again the observation of Kenton et al. [53] is of interest who did not find an improvement in NEC-associated mortality with an increasing number or volume of platelet transfusions.

6. Recommendations for treatment with platelet transfusions

Platelet transfusions are commonly used in preterm infants with NT at different threshold values. In 2018, a milestone study prospectively investigated whether platelet transfusion should be given at platelet-count thresholds of 50,000/μL (high-threshold group) or 25,000/μL (low-threshold group) [52]. In this multicenter trial, preterm infants born at less than 34 weeks of gestation in whom severe thrombocytopenia was diagnosed were included and randomly assigned to high or low platelet transfusion groups. The primary outcome was death or new major bleeding within 28 days after randomization. Of 660 infants (median birth weight, 740 g; and median gestational age, 26.6 weeks), 90% of the infants (296 of 328 infants) of the high-threshold group received at least one platelet transfusion, as compared with 53% (177 of 331 infants) in the low-threshold group. A new major bleeding episode or death happened in 26% of the infants in the high-threshold group and in 19% in the low-threshold group (odds ratio, 1.57; 95% confidence interval 1.06–2.32; p = 0.02). Most exciting, there was no significant difference between corresponding rates of serious adverse events (25 vs. 22%) [51].

An overview of recommendations at different thresholds of thrombocyte counts is given in **Table 2**.

Cut-off value thrombocyte counts	Recommendation	Authors
<20 × 10^9/L	All neonates (prophylactic)	Blanchette et al. [54, 55]; Gibson et al. [56]; Chakravorty et al. [57]; Carr et al. [58]
20–29 × 10^9/L	Non-bleeding term and preterm infant	Blanchette et al. [55]; Roberts et al. [59]; Calhoun et al. [60]; Sola-Visner et al. [61]; Murray [62]; Gibson et al. [56]; Carr et al. [58]; Sparger et al. [63]
	Bleeding neonates	Murray et al. [62]; Roberts et al. [4, 44]; Chakravorty et al. [57]
30–49 × 10^9/L	Preterm infants (unstable neonate, first week of life, surgery or invasive procedures)	Blanchette et al. [54, 55]; Roberts et al. [59]; Sola-Visner et al. [61]; Calhoun et al. [60]; Murray [62]; Roberts et al. [4]; Sparger et al. [63]
	Bleeding neonates	Blanchette et al. [54]; Roberts et al. [44, 59]; Gibson et al. [56]; Chakravorty et al. [57]
50–99 × 10^9/L	Non-bleeding sick preterm	Blanchette et al. [54, 55]; Roberts et al. [59]
	Bleeding neonates	Roberts et al. [4, 44, 59]; Murray et al. [62]; Sola-Visner et al. [61]; Sparger et al. [63]
100–150 × 10^9/L	No recommendations	—

Table 2.
Recommendations for platelet transfusions depending on different thrombocyte counts.

7. Conclusions

In conclusion, neonatal thrombocytopenia is a common problem at the neonatal intensive care unit. In most cases, it is a mild to moderate, self-limited entity. In severe cases, immune-mediated disease has to be suspected warranting prompt diagnosis and careful management. The threshold for platelet transfusions better should be at thrombocyte counts of 25–30,000/μL due to recent data reporting on a reduced mortality rate using a more restrictive transfusion regimen.

Acknowledgements

I want to thank my daughter Elisabeth who published our own data in 2018 and inspired me to write this chapter.

Conflict of interest

The author declares no conflict of interest.

Author details

Bernhard Resch
Division of Neonatology, Department of Pediatrics, Medical University of Graz, Graz, Austria

*Address all correspondence to: bernhard.resch@medunigraz.at

References

[1] Resch E, Hinkas O, Urlesberger B, Resch B. Neonatal thrombocytopenia—Causes and outcomes following platelet transfusions. European Journal of Pediatrics. 2018;**177**:1045-1052

[2] Wiedmeier SE, Henry E, Sola-Visner MC, Christensen RD. Platelet reference ranges for neonates, defined using data from over 47,000 patients in a multihospital healthcare system. Journal of Perinatology. 2009;**29**:130-136

[3] Sola MC, Del Vecchio A, Rimsza LM. Evaluation and treatment of thrombocytopenia in the neonatal intensive care unit. Clinics in Perinatology. 2000;**27**:655-679

[4] Roberts I, Murray NA. Neonatal thrombocytopenia: Causes and management. Archives of Disease in Childhood. Fetal and Neonatal Edition. 2003;**88**:F359-F364

[5] Del Vecchio A. Evaluation and management of thrombocytopenic neonates in the intensive care unit. Early Human Development. 2014;**90**(S2):551-555

[6] Dahmane Ayadi I, Ben Hamida E, Youssef A, Sdiri Y, Marrakchi Z. Prevalence and outcomes of thrombocytopenia in a neonatal intensive care unit. La Tunisie Médicale. 2016;**94**(4):305-308

[7] Ulusoy E, Tüfekçi O, Duman N, Kumral A, Irken G, Oren H. Thrombocytopenia in neonates: Causes and outcomes. Annals of Hematology. 2013;**92**:961-967

[8] Von Lindern J, Van den Bruele T, Lopriore E, Walther FJ. Thrombocytopenia in neonates and the risk of intraventricular hemorrhage: A retrospective cohort study. BMC Pediatrics. 2011;**11**:16

[9] Christensen RD, Baer VL, Henry E, Snow GL, Butler A, Sola-Visner MC. Thrombocytopenia in small-for-gestational-age infants. Pediatrics. 2015;**136**:e361-e370

[10] Fustolo-Gunnink SF, Vlug RD, Smits-Wintjens VE, Heckman EJ, Te Pas AB, Fijnvandraat K, et al. Early-onset thrombocytopenia in small-for-gestational-age neonates: A retrospective cohort study. PLoS One. 2016;**11**:e0154853

[11] Eslami Z, Lookzadeh MH, Noorishadkam M, Hashemi A, Ghilian R, PirDehghan A. Thrombocytopenia and associated factors in neonates admitted to NIC during years 2010_2011. Iranian Journal of Pediatric Hematology and Oncology. 2013;**3**:205-215

[12] Venkatesh V, Curley AE, Clarke P, Watts T, Stanworth SJ. Do we know when to treat neonatal thrombocytopaenia? Archives of Disease in Childhood. Fetal and Neonatal Edition. 2013;**98**:F380-F382

[13] Sola-Visner M, Saxonhouse MA, Brown RE. Neonatal thrombocytopenia: What we do and don't know. Early Human Development. 2008;**84**:499-506

[14] Jeremiah Z, Oburu J, Ruggeri M. Pattern and prevalence of neonatal thrombocytopenia in Port Harcourt Nijeria. Pathology and Laboratory Medicine International. 2010;**2**:27-31

[15] Naguri MH, Mubaeie C, Mathaei H. Incidence of thrombocytopenia in the NICU. Medical Journal, Armed Forces India. 2011;**67**:234-236

[16] Mehta P, Vasa R, Neumann L, Karpatkin M. Thrombocytopenia in the high-risk infant. The Journal of Pediatrics. 1980;**97**:791-794

[17] Christensen RD, Henry E, Wiedmeier SE, Stoddard RA, Sola-Visner MC, Lambert DK, et al. Thrombocytopenia among extremely low birth weight neonates: Data from a multihospital healthcare system. Journal of Perinatology. 2006;**26**:348-353

[18] Beiner ME, Simchen MJ, Sivan E, Chetrit A, Kuint J, Schiff E. Risk factors for neonatal thrombocytopenia in preterm infants. American Journal of Perinatology. 2003;**20**:49-54

[19] Sola-Visner M, Sallmon H, Brown R. New insights into the mechanisms of nonimmune thrombocytopenia in neonates. Seminars in Perinatology. 2009;**33**:43

[20] Arnold DA. Drug-induced immune thrombocytopenia. UpToDate. 2020. Available from: https://www.uptodate.com/contents/drug-induced-immune-thrombocytopenia

[21] Reese JA, Nguyen LP, Buchanan GR, et al. Drug-induced thrombocytopenia in children. Pediatric Blood & Cancer. 2013;**60**:1975

[22] Kaushansky K. The molecular mechanisms that control thrombopoiesis. The Journal of Clinical Investigation. 2005;**115**:3339-3347

[23] Murray NA, Roberts IA. Circulating megakaryocytes and their progenitors in early thrombocytopenia in preterm neonates. Pediatric Research. 1996;**40**:112e9

[24] Saxonhouse MA, Rimsza LM, Stevens G, et al. Effects of hypoxia on megakaryocyte progenitors Ob obtained from the umbilical cord blood of term and preterm neonates. Biology of the Neonate. 2006;**89**:104-108

[25] Lambert MP, Rauova L, Bailey M, et al. Platelet factor 4 is a negative autocrine in vivo regulator of megakaryopoiesis: Clinical and therapeutic implications. Blood. 2007;**110**:1153-1160

[26] Cole JL, Marzec UM, Gunthel CJ, et al. Ineffective platelet production in thrombocytopenic human immunodeficiency virus-infected patients. Blood. 1998;**91**:3239-3246

[27] Tighe P, Rimsza LM, Christensen RD, et al. Severe thrombocytopenia in a neonate with congenital HIV infection. The Journal of Pediatrics. 2005;**146**:408-413

[28] Harrington WJ, Sprague CC, Minnich V, Moore CV, Aulvin RC, Dubach R. Immunologic mechanisms in idiopathic and neonatal thrombocytopenic purpura. Annals of Internal Medicine. 1953;**38**:433-469

[29] Shulman NR, Aster RH, Pearson HA, Hiller MC. Immunoreactions involving platelet. VI. Reactions of maternal isoantibodies responsible for neonatal purpura. Differentiation of a second platelet antigen system. The Journal of Clinical Investigation. 1962;**41**:1059-1069

[30] Van Loghem J Jr, Dorfmeijer H, Van Hart M, Schreuder F. Serological and genetical studies on a platelet antigen (Zwa). Vox Sanguinis. 1959;**4**:161-169

[31] Peterson JA, McFarland JG, Curtis BR, Aster RH. Neonatal alloimmune thrombocytopenia: Pathogenesis, diagnosis and management. British Journal of Haematology. 2013;**161**:3-14

[32] Williamson LM, Hackett G, Rennie J, Palmer CR, Maciver C, Hadfield R, et al. The natural history of fetomaternal alloimmunization to the platelet-specific antigen HPA-1a (PlA1, Zwa) as determined by antenatal screening. Blood. 1998;**92**:2280-2287

[33] Blanchette VS, Chen L, de Friedberg ZS, Hogan VA, Trudel E,

Décary F. Alloimmunization to the PlA1 platelet antigen: Results of a prospective study. British Journal of Haematology. 1990;**74**:209-215

[34] Kjeldsen-Kragh J, Killie MK, Tomter G, Golebiowska E, Randen I, Hauge R, et al. A screening and intervention program aimed to reduce mortality and serious morbidity associated with severe neonatal alloimmune thrombocytopenia. Blood. 2007;**110**:833-839

[35] Bussel JB, Sola-Visner M. Current approaches to the evaluation and management of the fetus and neonate with immune thrombocytopenia. Seminars in Perinatology. 2009;**33**:35-42

[36] Kamphuis MM, Paridaans NP, Porcelijn L, Lopriore E, Oepkes D. Incidence and consequences of neonatal alloimmune thrombocytopenia: A systematic review. Pediatrics. 2014;**133**:715-721

[37] Winkelhorst D, Oepkes D, Lopriore E. Fetal and neonatal alloimmune thrombocytopenia: Evidence based antenatal and postnatal management strategies. Expert Review of Hematology. 2017;**10**:729-737

[38] Kamphuis MM, Oepkes D. Fetal and neonatal alloimmune thrombocytopenia: Prenatal interventions. Prenatal Diagnosis. 2011;**31**:712-719

[39] Zdravic D, Yougbare I, Vadasz B, Li C, Marshall AH, Chen P, et al. Fetal and neonatal alloimmune thrombocytopenia. Seminars in Fetal & Neonatal Medicine. 2016;**21**:19-27

[40] Paidas MJ. Fetal and neonatal alloimmune thrombocytopenia: Parental evaluation and pregnancy management. Uptodate; 2020

[41] Rayment R, Brunskill SJ, Soothill PW, Roberts DJ, Bussel JB, Murphy MF. Antenatal interventions for fetomaternal alloimmune thrombocytopenia. Cochrane Database of Systematic Reviews. 2011;**5**:CD004226

[42] Baer VL, Lambert DK, Henry E, Christensen RD. Severe thrombocytopenia in the NICU. Pediatrics. 2009;**124**:e1095-e1100

[43] de Alarcon PA. Neonatal Hematology. Cambridge: Cambridge, New York: University Press; 2005. p. 199

[44] Roberts I, Murray NA. Neonatal thrombocytopenia. Seminars in Fetal and Neonatal Medicine. 2008;**13**:256-264

[45] Sparger KA, Assmann SF, Granger S, Winston A, Christensen RD, Widness JA, et al. Platelet transfusion practices among very-low-birth-weight infants. JAMA Pediatrics. 2016;**170**:687-694

[46] Christensen RD. Platelet transfusion in the neonatal intensive care unit: Benefits, risks, alternatives. Neonatology. 2011;**100**:311-318

[47] Saxonhouse MA, Manco-Johnson MJ. The evaluation and management of neonatal coagulation disorders. Seminars in Perinatology. 2009;**33**:52-65

[48] Deaver JE, Leppert PC, Zaroulis CG. Neonatal alloimmune thrombocytopenic purpura. American Journal of Perinatology. 1986;**3**:127-131

[49] Murray NA, Howarth LJ, McCloy MP, Letsky EA, Roberts IA. Platelet transfusion in the management of severe thrombocytopenia in neonatal intensive care unit patients. Transfusion Medicine. 2002;**12**:35-41

[50] Garcia MG, Duenas E, Sola MC, Hutson AD, Theriaque D, Christensen RD. Epidemiologic and

outcome studies of patients who received platelet transfusions in the neonatal intensive care unit. Journal of Perinatology. 2001;**21**:415e20

[51] Del Vecchio A, Sola MC, Theriaque DW, Hutson AD, Kao KJ, Wright D, et al. Platelet transfusions in the neonatal intensive care unit: Factors predicting which patients will require multiple transfusions. Transfusion. 2001;**41**:803-808

[52] Curley A, Stanworth SJ, Willoughby K, Fustolo-Gunnink SF, Venkatesh V, Hudson C, et al. PlaNeT2 MATISSE collaborators. Randomized trial of platelet-transfusion thresholds in neonates. The New England Journal of Medicine. 2019;**380**:242-251

[53] Kenton AB, Hegemier S, Smith EO, O'Donovan DJ, Brandt ML, Cass DL, et al. Platelet transfusions in infants with necrotizing enterocolitis do not lower mortality but may increase morbidity. Journal of Perinatology. 2005;**25**:173e7

[54] Blanchette VS, Hume HA, Levy GJ, Luban NL, Strauss RG. Guidelines for auditing pediatric blood transfusion practices. American Journal of Diseases of Children. 1991;**145**:787-796

[55] Blanchette VS, Kühne T, Hume H, Hellmann J. Platelet transfusion therapy in newborn infants. Transfusion Medicine Reviews. 1995;**9**:215-230

[56] Gibson BE, Todd A, Roberts I, Pamphilon D, Rodeck C, Bolton-Maggs P, et al. Transfusion guidelines for neonates and older children. British Journal of Haematology. 2004;**124**:433-453

[57] Chakravorty S, Roberts I. How I manage neonatal thrombocytopenia. British Journal of Haematology. 2012;**156**:155-162

[58] Carr R, Kelly AM, Williamson LM. Neonatal thrombocytopenia and platelet transfusion - a UK perspective. Neonatology. 2015;**107**:1-7

[59] Roberts IA, Murray NA. Management of thrombocytopenia in neonates. British Journal of Haematology. 1999;**105**:864-870

[60] Calhoun DA, Christensen RD, Edstrom CS, Juul SE, Ohls RK, Schibler KR, et al. Consistent approaches to procedures and practices in neonatal hematology. Clinics in Perinatology. 2000;**27**:733-753

[61] Sola-Visner M, Bercovitz RS. Neonatal platelet transfusions and future areas of research. Transfusion Medicine Reviews. 2016;**30**:183-188

[62] Murray NA. Evaluation and treatment of thrombocytopenia in the neonatal intensive care unit. Acta Paediatrica. 2002;**91**(Suppl 438):74-81

[63] Sparger K, Deschmann E, Sola-Visner M. Platelet transfusions in the NICU. Clinics in Perinatology. 2015;**42**:613-623

Section 3

Platelet Application

Chapter 7

Platelet Rich Fibrin (PRF) Application in Oral Surgery

Alper Saglanmak, Caglar Cinar and Alper Gultekin

Abstract

Platelet rich fibrin (PRF) is an autologous biological product which becomes popular day by day and available in a wide variety of fields in medicine. Platelet concentrates which are introduced at the early 90s have evolved over the years. The use such autologous materials have become trendy in recent years to encounter demanding expectations of patients, improve treatment success and maximize patient comfort. Despite its increasing use in dentistry and oral surgery, the most indications and effects are still being discussed. PRF is easily accepted by patients because of its low cost, easy to receive, low donor morbidity, low postoperative complication and infection rate. This biomaterial may be a solution for patients who have strong negative beliefs about the use of allografts and xenografts or who are afraid of complications during the grafting procedure. The objectives of these technologies are to use their synergistic effect to improve the hard and soft tissue regeneration. PRF in oral surgery are used for alveolar bone reconstruction, dental implant surgery, sinus augmentation, socket preservation, osteonecrosis, oroantral fistula closure, struggling with oral ulcers, preventing swelling and edema constitution. This chapter aims to review the clinical applications of platelets in oral surgery and the role of molecular components in tissue healing.

Keywords: platelet rich fibrin, oral surgery, tissue healing, dental implant

1. Introduction

In recent years, the question how to increase patient comfort after surgical interventions became the main topic of oral surgical applications. In addition to minimally invasive surgical techniques, extra procedures performed during or after surgery are aimed to reduce postoperative morbidity. As a result of various researches in recent years, the use of platelet concentrates give rise to improve patient comfort and enhance healing after the operation.

Surgical techniques to gain bone and soft tissue can be difficult and associated with higher morbidity. Although they are technique delicate, they just considered as gold-standard, because of their capacity in healing enhancement [1]. However, alternative autologous blood derivatives such as platelet rich fibrin (PRF) are becoming a current issue with its easy use and effectiveness.

PRF is an autologous product acquired from the patient's own blood and enters dental field as a second- generation platelet concentrate under the name PRF [Platelet Rich Fibrin] by Choukroun. Although they are known by different names according to the centrifugation time (A-PRF, L-PRF, I- PRF, P-PRF) their main rationale is the same. They are increasing the healing capability of the tissue by releasing growth factors from platelet granules. These factors are essential for inflammation process and they have positive effect on healing enhancing. Since it is an autologous product, it does not cause allergic reactions, it can be prepared rapidly and easily, there is no risk of disease transfer and no risk of donor site morbidity. The main advantages are controlling inflammation and suppressing infection by leukocyte and cytokine secretion [2] (**Figures 1** and **2**).

In order to get in depth information about the supportive effect of PRF one should know about the healing pattern of injured tissue. There are four sequential phases of wound healing: Hemostasis, inflammatory, proliferative and remodeling phase [3]. At hemostasis phase, platelets are essential for blood clot formation and PRF with rich platelet granules is promotive to accommodate a strong fibrin network. This blood clot serves as a reservoir which allows cell migration, adhesion and proliferation. The impact of this fibrin matrix proceeds throughout the whole healing process. Inflammatory phase starts with the injury and takes 5–7 days approximately. During this phase, platelets are releasing various growth factors to the injured site that migrate inflammatory cells (Lymphocytes, macrophages and neutrophils). These factors are PDGF (Platelet-derived growth factor), VEGF (Vascular endothelial growth factor), TGF B (Transforming growth factor) and pro-inflammatory cytokines such as interleukins (IL-1, IL-6, IL-8) and tumor necrosis factor alpha (TNF-α), whose roles are enhancing angiogenesis and tissue healing. Within the comprising of new blood vessels with angiogenesis, acidic and hypoxic environment change. In the proliferative phase, MSCs (Mesenchymal stem cells) releasing from newly formed blood vessels, BMP's (Bone morphogenic protein) and TGF-β are playing an important role in MSCs organism. MSC's role is inducing osteoblast differentiation. The last but not the least, remodeling phase is characterized as maturation process. Within this process, vascularity ratio and collagen deposition decreases and mineral deposition increases with the replacement of woven bone into lamellar one [4–6].

Fibrin forms a matrix for the migration of cells such as fibroblasts and endothelial cells, which are crucial in angiogenesis and new tissue formation. PRF is a strong fibrin matrix structure, platelets and leukocytes attach on it and activate degranulated growth factors with the consequence of releasing cytokines. It has been suggested that PRF, a natural fibrin network, can protect the growth factors containing in its own structure from proteolysis. Thus growth factors may maintain their activity for a long time and stimulate tissue regeneration [7].

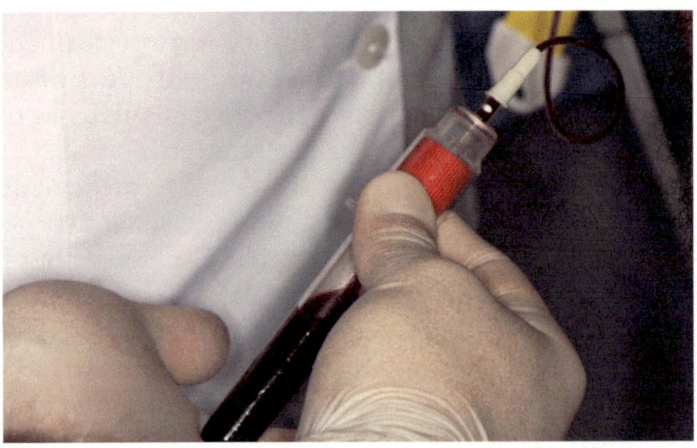

Figure 1.
Bloodletting of a patient.

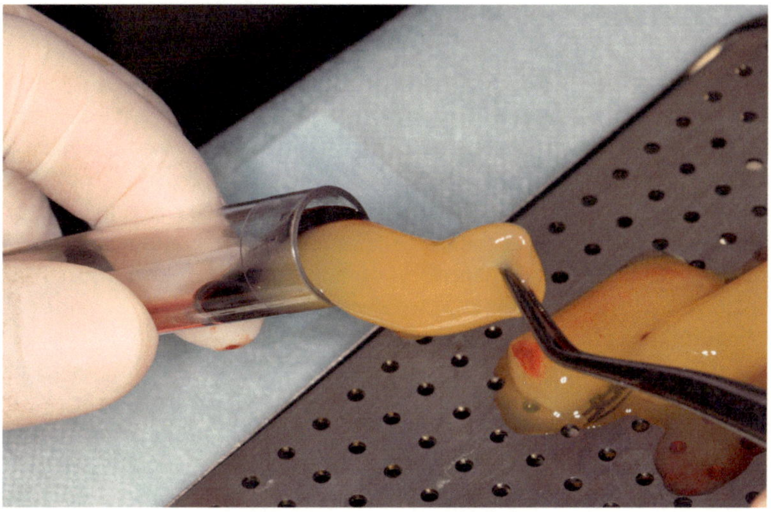

Figure 2.
PRF membrane preparation.

1.1 Types of PRF

The main purpose of using PRF is to release the rich content of alpha granules of platelets into the environment for therapeutic purposes. In addition to the basic functions of platelets, the contents of alpha and dense granules are very important for different processes such as inflammation and angiogenesis [8]. The main differences between PRF types are their centrifuge speed.

- **Advanced platelet rich fibrin (A-PRF)**: It is obtained with longer centrifugation time and lower rpm. Thus, more neutrophilic granulocytes are present in the distal region of the clot. Neutrophilic granulocytes contribute to monocyte differentiation in macrophages (**Figure 3**).

- **Pure platelet rich fibrin (P-PRF):** After the first centrifuge (6 min high speed), transferring the buffy coat and PPP (Platelet poor plasma) to the second tube, which contains $CaCl_2$. After the second centrifuge starts and takes 15 min long, stable platelet-fibrin takes place. The authenticity of this method is the presence of separation gel at the first tube.

- **Leukocyte and platelet rich fibrin (L-PRF):** It is very simple and cheap method. Blood samples are taken into glass tubes without any anticoagulant and centrifuged immediately at low speed. It is formed three different layers with acellular plasma, platelet-rich fibrin and erythrocyte layer at the bottom, respectively. Thrombocyte rich fibrin matrix is very powerful and autologous biomaterial can be used in different fields in oral surgery.

- **Injectable platelet rich fibrin (I-PRF):** Blood involves high number of leukocytes. However coagulation occurs within few minutes after the centrifugation has finished. The use of I-PRF [Injectable] is at an early stage. But the results are very promising in terms of increasing vascularity and soft tissue healing.

- **Liquid platelet rich fibrin (Liquid-PRF):** Liquid-PRF was defined with low-speed centrifugation (LSC), which allows forming of a liquid-PRF formula of fibrinogen and thrombin rather than its conversion to fibrin [2].

1.2 PRF in post extraction

Tooth extraction has various adverse effects such as pain, bleeding, swelling, infection etc. Wound healing in the tooth extraction is characterized by bone loss as a natural process. Furthermore; extraction will result recession around adjacent teeth and hinders the functional and esthetic prosthetic rehabilitation. PRF have been shown to play an important role in tissue healing with the releasing growth factors from alpha granules, regulate cellular events such as cell adhesion, migration, proliferation, differentiation and extracellular matrix deposition. Major changes occurred within the first year following extraction, but a major part of bone resorption takes place only within 3 months [9, 10].

The rationale behind the enhancement of PRF in socket healing is very slowly polymerizing, cell migration and fibrin network capable of proliferation. During remodeling of the fibrin network many important growth factors from activated platelets and the release of the matrix glycoproteins. This biochemical structure gives rise to the tissue regeneration (**Figures 4** and **5**).

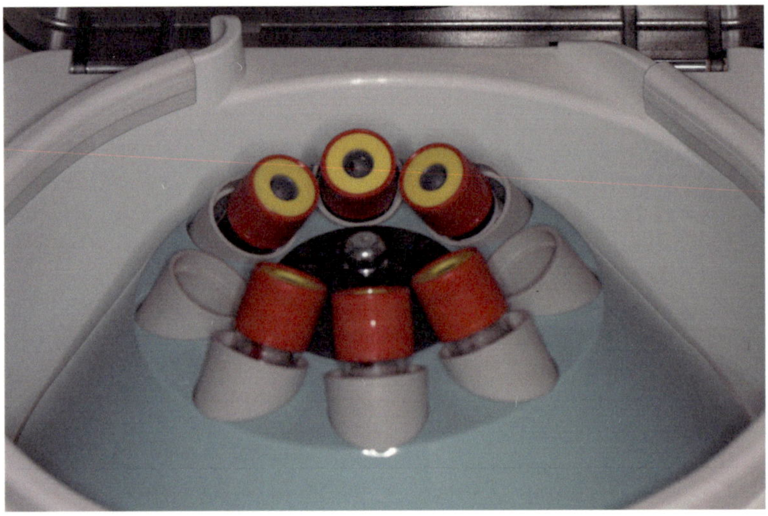

Figure 3.
A-PRF centrifuge device.

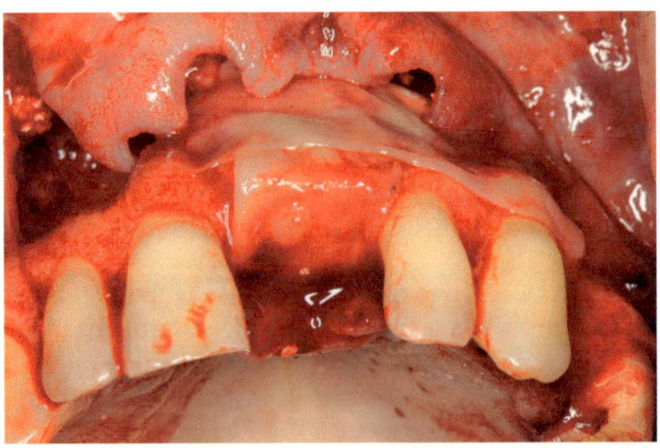

Figure 4.
PRF membrane utilization in guided bone regeneration.

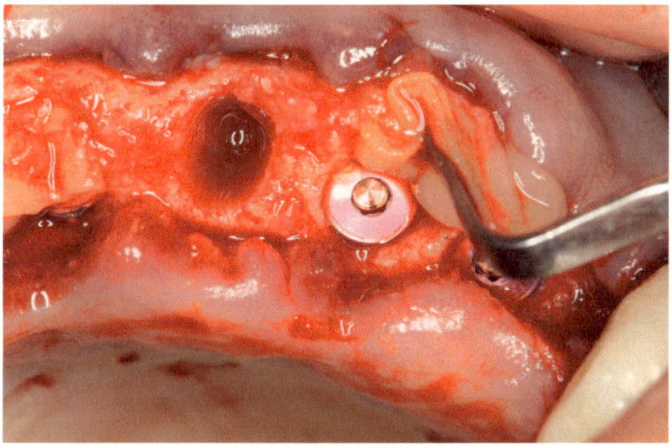

Figure 5.
PRF membrane application in immediate implantation for filling the buccal gap.

On the other hand the role of PRF in soft tissue healing has been shown at well-designed meta-analyses. In fact there are no significant differences in alveolar osteitis, acute inflammation or alveolar infection following tooth extraction. New bone gain and bone remodeling topic is also contradictory. However it was concluded that PRF is good at decreasing swelling, edema, pain and trismus following tooth extraction [9–11].

1.3 PRF for maintaining swelling, edema and pain

Modern studies on clinical research showed that PRF aim to increase not only the success of the treatments but also the patient comfort. In this sense, the use of autologous products is advantageous such as high acceptability, low risk of disease transfer, low morbidity and low cost. PRF plays a crucial role in tissue repair. Their alpha granules include many substances, plenty of growth factors with significant effects on the inflammatory and proliferative resident cells at the site of injury, like mesenchymal stem cells, fibroblasts, chondrocytes and osteoblasts. This potential may be increased by the concentration of the platelets. Certain in-vitro and in-vivo studies have shown that the use of PRF is significantly advantageous in terms of cell migration. Most of clinical studies found the use of PRF positive at wound healing. According to the results of clinical studies, the use of PRF provides an advantage in soft tissue healing, reducing swelling and trismus and increasing patient comfort. However evidences on maintaining pain is scarce and pain usually adhere on early formation of soft tissue healing [12].

1.4 PRF in periodontal treatment

PRF can also be used in regenerative periodontal therapy to enhance hard and soft tissue wound healing and promote periodontal tissue regeneration. Various studies have shown the favorable benefit of using PRF as an adjunct to traditional periodontal surgical techniques. These studies all exhibit improved clinical outcomes regarding key clinical parameters such as clinical attachment level and pocket depth with the use of PRF when compared to conventional techniques applied alone.

The rationale behind this benefit is believed to lie in the differentiation and proliferation inducing abilities of PRF. The rich source of bioactive cells within PRF itself stimulate the local environment and regulate the inflammation process, thereby enhancing periodontal wound healing and reducing postoperative discomfort. In addition to these benefits the PRF sample also inherently supplies growth factors, releasing them slowly into the wound for 7–14 days. Other obvious benefits include graft stabilization. Furthermore, a possible antimicrobial effect of L-PRF is also present [13].

Another reason for PRF being favored in periodontal therapy is its multipurpose nature. The centrifuged buffy coat can be used alone in defects, combined with particulate graft materials or as a thin membrane covering in regeneration techniques. Studies also show that PRF can be used as an alternative to connective tissue grafts (CTG) in periodontal plastic surgery owing to its cellular contents. Additionally, the many benefits of PRF use in periodontal therapy also include its graft stabilization, wound sealing and hemostatic abilities. Evidently, along with its favorable biologic outcomes and low-cost PRF seems to be almost ideally suited for various periodontal purposes [13, 14].

The performance of PRF in different periodontal surgery indications was measured and PRF was found to perform superiorly when compared to conventional perio-plastic surgeries applied alone. Its use in intra-bony defects and furcation defects have proved beneficial in reducing pocket depth values, clinical attachment level gains and bone fill percentages. Improved outcomes in intrabony defects were obtained when used alone or in conjunction with other biomaterials. In furcation defects also, traditional flap surgeries tended to perform better when complemented with PRF. Coronally Advanced Flap (CAF) procedures showed improved results when accompanied with either CTGs or PRF membranes. Compared to each other however, these two materials seemed to perform similarly. Therefore, it can only be said that PRF can be considered a suitable alternative to CTGs in periodontal plastic surgery [14, 15].

1.5 PRF in sinus lifting

Implant rehabilitation success is highly related with sufficient bone volume and density. The posterior maxilla represents a challenging and unique area for successful dental implant rehabilitation because of its relatively deficient bone volume and poor bone quality caused by alveolar bone resorption and maxillary sinus pneumatization. Rehabilitation of posterior maxillary bone volume has been successes by different procedures, such as Le Fort I osteotomies, onlay grafts and sinus lifts [16, 17]. Maxillary sinus floor elevation is considered one of the most successful procedures that can be performed using different grafting materials, such as autogenous, xenograft, allograft, alloplast and PRF [18, 19].

Autogenous bone with osteogenic, osteoinductive and osteoconductive properties is still considered to be the gold standard. However, grafting with autogenous bone is associated with donor site morbidity, extended duration of surgical procedures and the volume of bone graft harvested may be insufficient for the requirements. Biomaterials, thus, are promising substitutes for autogenous bone grafts in maxillary sinus augmentation. Osteoconductive properties of these biomaterials have been shown in clinical studies with satisfactory clinical outcomes.

On the other hand, these bone graft materials demonstrate lack of osteogenic and osteoinductive potential with distinct osteogenic capacity and bone formation. Moreover, some disadvantages, mainly related to a limited availability, prolonged healing time and impact on host responses can appear when using these bone substitutes. To overcome these problems, new substances with osteoinductive properties, such as platelet-rich fibrin (PRF) was recently introduced as replacement or additional materials in sinus augmentation procedures [20].

The biologic mediators have osteoinductive properties and they are considered to accelerate the formation of new bone and to reduce the time interval. The strengths of PRF comes from promoting the vascularization of bone tissue, reducing tissue inflammation, improving scaffold mechanics and accelerating new bone formation [20]. Newly, researchers have paid greater attention to the success of PRF application in maxillary sinus lifting procedures, but no consensus has been reached. Some researchers have reported positive effects of PRF application in sinus augmentation procedures.

Platelet concentrates have been used to accelerate bone generation and improve healing by releasing growth factors such as transforming growth factor β1 and β2, platelet-derived growth factor and vascular endothelial growth factor, which are able to induce angiogenesis and activate cell proliferation.

In the literature there are some different application techniques for PRF in the sinus augmentation such as PRF as a sole grafting material, PRF with allografts or PRF with xenografts. All of these techniques have variable clinical, radiographic and histologic and histomorphometric outcomes.

Mazor et al. [21] and Simonpieri et al. [22] performed sinus lift by using lateral approach and PRF was used as a sole grafting material and implants were applied immediately to serve as tent pegs. During the healing period there were no complications. A 100% survival rate was observed in total of 57 sinus lift procedures and 110 implants during the follow-up period (2 years). Radiographic examination was performed by CT scan or panoramic radiographs about 6 months after the sinus augmentation to examine the bone volume, where the average bone gain was 9.8 mm. Histologic and histomorphometric examination accomplished by Mazor et al. showed that dense collagen matrix, easily identified osteocytes and osteoblasts in the lacunae and well-organized and vital bone with structured trabeculae with more than %30 bone matrix.

Choukroun et al. [23] performed sinus augmentation with PRF in combination with demineralized freeze-dried bone allograft (DFDBA). They found the rate of vital bone/inert bone%20 both in test and control group but with a reduced healing time at PRF group.

Zhang et al. [24] applied the PRF/xenograft mixture for the test group and xenograft as a sole graft material for the control group. They found no statistically significant difference between the two groups.

In light of this information, although there is not a consensus statement about the effect of PRF as a grafting material at sinus lifting procedure, still it is a good alternative material with its osteoinductive properties to enhance hard tissue healing.

1.6 PRF for preserving bone around implants

Marginal bone loss is an inevitable process which starts immediately following implant placement. There have been done plenty of studies since decades to minimize it. Previous studies about preserving bone around implants, has focused on soft tissue thickness and it was hypothesized, adequate soft tissue volume around implants has a positive effect in preserving marginal bone and PRF is perfect material to augment soft tissue. We know PRF is a good autologous material to enhance soft tissue healing with its growth factors including VEGF, PRGF, etc. However researches about PRF usage to augment hard tissue have contradictory results and there is need to do further detailed randomized controlled clinical studies to know about the effect of PRF preserving marginal bone [25] (**Figures 6** and 7).

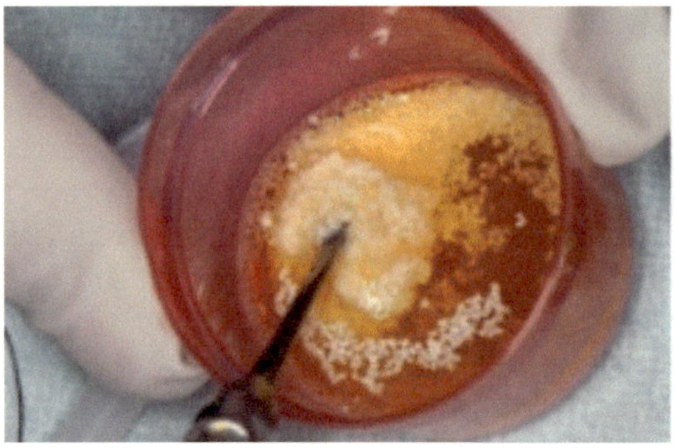

Figure 6.
Liquid PRF combined with MPM (Mineralized Plasmatic Matrix) and B-TCP graft material for sinus lifting.

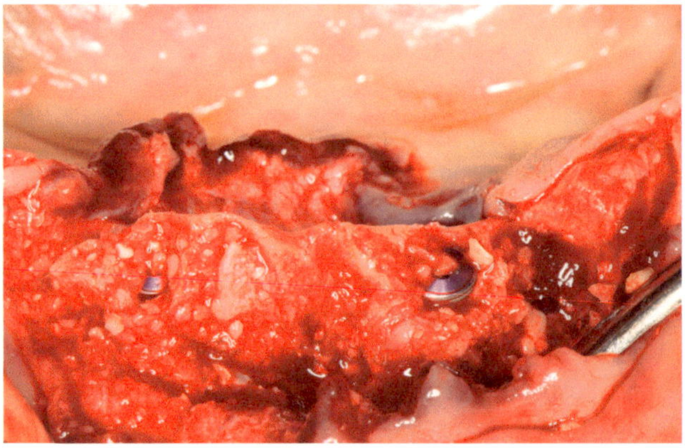

Figure 7.
PRF pieces combined with xenograft in guided bone regeneration.

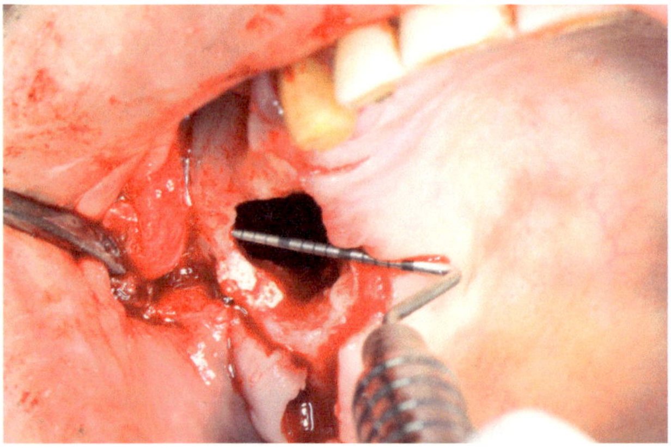

Figure 8.
Large sinus membrane perforation.

1.7 PRF in rare clinical scenarios

PRF could also be beneficial with growth factors including in rare clinical scenarios such as cyst treatment, sinus membrane perforations, oroantral fistulae closure and osteonecrosis.

Oroantral fistula (OAF) constitution is defined a pathological way between maxillary sinus and oral cavity. It is an unnatural epithelial connection filled with granulation tissue or polypoid extension of sinus membrane. It can either come into existence spontaneously following a large maxillary cyst or tumor or as iatrogenic after tooth extraction or dental implant surgery [26]. An OAF is highly iatrogenic and depending on the perforation of sinus membrane during surgical interventions at maxilla. Either this or that way the treatment of perforated sinus membrane is bringing the tissue free from infection, cleaning the epithelium and repairing the membrane (**Figure 8**).

There are plenty of methods maintaining with OAF. PRF is one of them which is recently introduced. The technique is as following; PRF clots obtained by centrifugation should be isolated from PPP (Platelet Poor Plasma) and red blood cells, prepared as thin membranes and applied perforated area layer by layer. The researches about PRF in OAF closure conclude that, wound healing is faster and there was an increase in soft tissue thickness during healing. Due to its natural ingredients, there are no need to use additional materials, thus less donor site morbidity occurs.

Osteonecrosis is another rare clinical scenario which is defined as avascular bone area surrounding soft tissue on occasion. Various clinical and medical considerations can cause osteonecrosis. It can occur in consequence of bisphosphonates (including denosumab), medications or iatrogenic dental malpraxis (In proper use of $NaOH_2$, formaldehyde, devitalizing agents). Either this or that way the healing of necrotic bone takes 8 weeks at least in appropriate circumstances. Osteonecrosis can conclude either with a demarcation line or heal just as avascular necrosis. Clinician should choose the treatment modality according to the clinical situation.

The main factor of osteonecrosis is the disturbance of vascular blood supply. The management of patient with compromised healing in bone is controversial. Despite the conventional treatment modalities, curettage of necrotic bone, antibiotic usage, chlorhexidine glukonate; with more complex treatment modalities hyperbaric oxygen therapy, ozone and low dose laser; PRF utilization alone or with these treatment modalities takes place in recent years (**Figure 9**) [27].

The use of PRF in cyst depends on the same rationale with the enhancement of soft and hard tissue healing. Researches related to this topic conclude that using PRF as a graft material is beneficial for shortening healing time and increasing bone mineral density (**Figures 10–12**).

Since it has been discovered the synergetic effect of PRF in healing enhancement of covering oral mucosa, these platelet derivatives became even more important. PRF

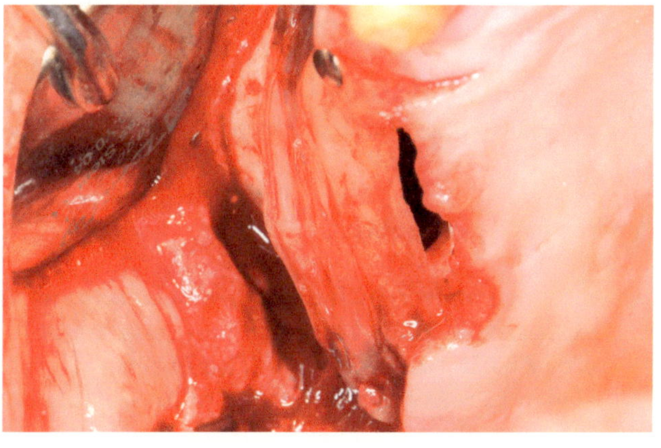

Figure 9.
Sinus membrane perforation closure with PRF membrane.

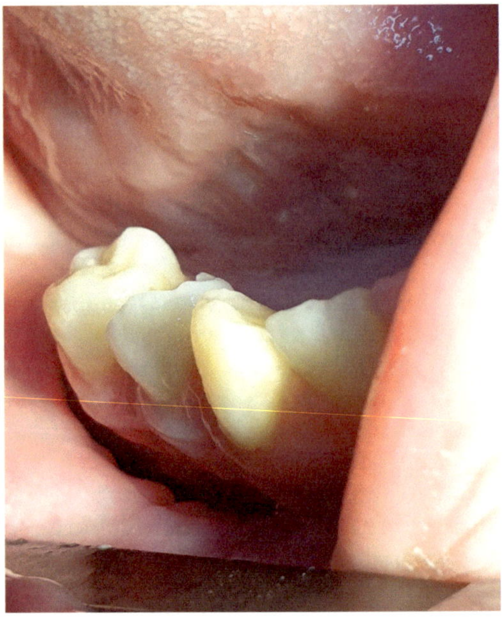

Figure 10.
2 weeks follow up after PRF application of avascular necrosis area in mandible.

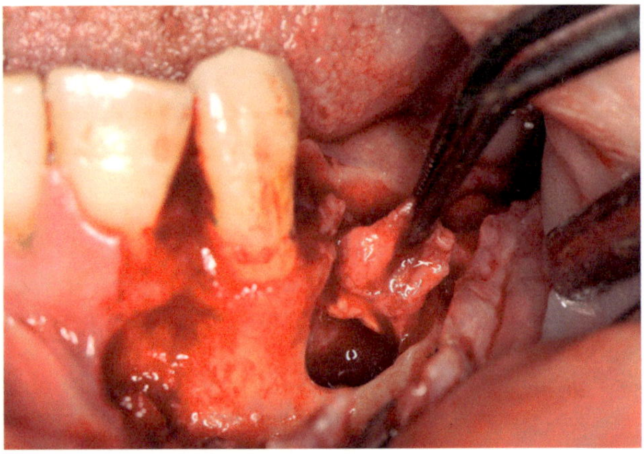

Figure 11.
Radicular cyst enucleation.

involves cytokines, chemokines, and antimicrobial derivatives with growth factors such as VEGF, which are crucial to support hard and soft tissue in order to heal (**Figure 13**).

2. Conclusion and future perspectives

PRF with its strong fibrin matrix, including growth factors and slow release, has a positive effect on wound healing. The most important factor for success in oral surgical procedures is early wound healing. This improvement will decrease the healing time and enhance relatively the healing of underlying bone. PRF's effect on decreasing pain, swelling and edema is evidenced based. Thus it is very promising material in applications mostly associated with soft tissue healing such as third molar extractions, oroantral fistula closure and alveolar cleft reconstructions.

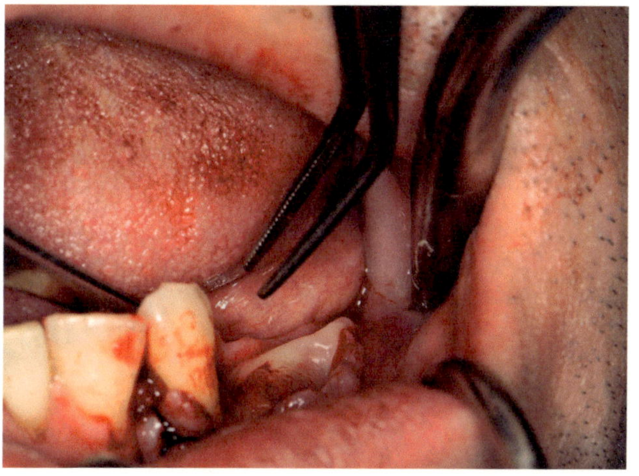

Figure 12.
PRF utilization after cyst enucleation.

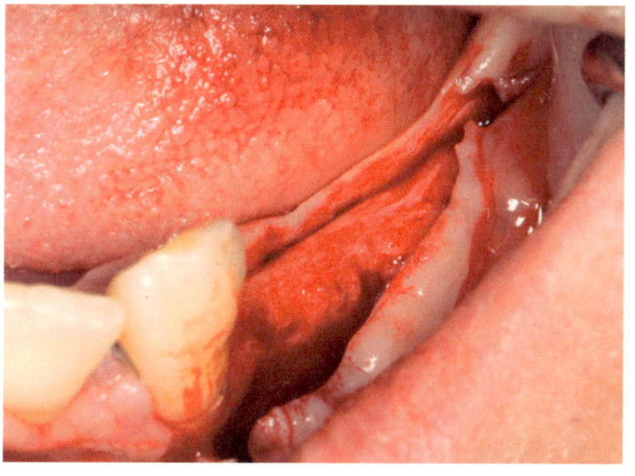

Figure 13.
Healing 2 months follow up after PRF augmentation.

However studies have sparsely mentioned about the positive effect of new bone formation at sinus lifting, periodontal and peri-implant bone preservation and alveolar bone augmentation. For future perspectives, with the use of new generations of PRF with increased growth factor capacity, combined with graft materials, PRF will appear in more areas in oral surgery applications.

Acknowledgements

Dt. Ali İsik from Istanbul University Dentistry Faculty Department of Oral Implantology, Istanbul, Turkey has valuable contributions to this book chapter.

Conflict of interest

The authors declare no conflict of interest.

Author details

Alper Saglanmak*, Caglar Cinar and Alper Gultekin
Dentistry Faculty, Department of Oral Implantology, Istanbul University, Istanbul, Turkey

*Address all correspondence to: alper.saglanmak@istanbul.edu.tr

References

[1] Sakkas A, Wilde F, Heufelder M, et al. Autogenous bone grafts in oral implantology—Is it still a "gold standard"? A consecutive review of 279 patients with 456 clinical procedures. International Journal of Implant Dentistry. 2017;**3**(23)

[2] Dohan DM, Choukroun J, Diss A, et al. Platelet-rich fibrin (PRF): A second-generation platelet concentrate. Part I: Technological concepts and evolution. Oral Surgery, Oral Medicine, Oral Pathology, Oral Radiology, and Endodontics. 2006;**101**(3):37-44

[3] Childs DR, Murthy AS. Overview of wound healing and management. Surgical Clinics of North America. 2017;**97**(1):189-207

[4] Gurtner GC, Werner S, Barrandon Y, Longaker MT. Wound repair and regeneration. Nature. 2008;**453**:314-321

[5] Salvi GE, Bosshardt DD, Lang NP, et al. Temporal sequence of hard and soft tissue healing around titanium dental implants. Periodontology 2000. 2015;**68**(1):135-152

[6] Araújo MG, Silva CO, Misawa M, et al. Alveolar socket healing: What can we learn? Periodontology 2000. 2015;**68**(1):122-134

[7] Gomes PS, Daugela P, Poskevicius L, et al. Molecular and cellular aspects of socket healing in the absence and presence of graft materials and autologous platelet concentrates: A focused review. Journal of Oral & Maxillofacial Research. 2019;**10**(3)

[8] Michelson AD. Chapter 6: The platelet proteome. In: Platelets. Amsterdam: Elsevier Inc.; 2019

[9] Del Fabbro M, Bucchi C, Lolato A, et al. Autologous platelet concentrates to improve post extraction outcomes. Journal of Oral and Maxillofacial Surgery. 2017;**75**:1601-1615

[10] Moraschini V, Barboza ESP. Effect of autologous platelet concentrates for alveolar socket preservation: A systematic review. International Journal of Oral and Maxillofacial Surgery. 2015;**44**:632-641

[11] Nanditha S, Balamanikandasrinivasan C, Senthilkumar M, et al. Appraising the diverse facets of platelet rich fibrin in surgery through a systematic review. International Journal of Surgery. 2017;**46**:186-194

[12] Shah N, Cains M. Autologous platelet concentrates to improve post extraction outcomes. Evidence-Based Dentistry. 2018;**19**:118-119

[13] Castro AB, Meschi N, Temmerman A, et al. Regenerative potential of leucocyte- and platelet-rich fibrin. Part A: Intra-bony defects, furcation defects and periodontal plastic surgery. A systematic review and meta-analysis. Journal of Clinical Periodontology. 2017;**44**(1):67-82

[14] Sculean A, Chappuis V, Cosgarea R. Coverage of mucosal recessions at dental implants. Periodontology 2000. 2017;**73**(1):134-140

[15] Krishnamurkar D, Mahendra J, Ari G, et al. A clinical and histological evaluation of platelet-rich fibrin and CGF for root coverage procedure using coronally advanced flap: A split-mouth design. Indian Journal of Dental Research. 2019;**30**(6):970-974

[16] Isaksson S. Evaluation of three bone grafting techniques for severely resorbed maxillae in conjunction with immediate endosseous implants. The International Journal of Oral & Maxillofacial Implants. 1994;**9**:679-688

[17] Chiapasco M, Casentini P, Zaniboni M. Bone augmentation procedures in implant dentistry. The International Journal of Oral & Maxillofacial Implants. 2009;**24**:237-259

[18] Yilmaz S, Karaca EO, Ipci SD, et al. Radiographic and histologic evaluation of platelet-rich plasma and bovinederived xenograft combination in bilateral sinus augmentation procedure. Platelets. 2013;**24**(4):308-315

[19] Kassolis JD, Reynolds MA. Evaluation of the adjunctive benefits of platelet-rich plasma in subantral sinus. The Journal of Craniofacial Surgery. 2005;**16**(2):280-287

[20] Zhao JH, Tsai CH, Chang YC. Clinical application of platelet-rich fibrin as the sole grafting material in maxillary sinus augmentation. Journal of the Formosan Medical Association. 2015;**114**(8):779-780

[21] Mazor Z, Horowitz RA, Del Corso M, et al. Sinus floor augmentation with simultaneous implant placement using Choukroun's platelet-rich fibrin as the sole grafting material: A radiologic and histologic study at 6 months. Journal of Periodontology. 2009;**80**:2056-2064

[22] Simonpieri A, Choukroun J, Del Corso M, et al. Simultaneous sinus-lift and implantation using microthreaded implants and leukocyte- and platelet-rich fibrin as sole grafting material: A six-year experience. Implant Dentistry. 2011;**20**:2-12

[23] Choukroun J, Diss A, Simonpieri A, et al. Platelet-rich fibrin (PRF): A second-generation platelet concentrate. Part V: Histologic evaluations of PRF effects on bone allograft maturation in sinus lift. Oral Surgery, Oral Medicine, Oral Pathology, Oral Radiology, and Endodontics. 2006;**101**:299-303

[24] Zhang Y, Tangl S, Huber CD, et al. Effects of Choukroun's platelet-rich fibrin on bone regeneration in combination with deproteinized bovine bone mineral in maxillary sinus augmentation: A histological and histomorphometric study. Journal of Cranio-Maxillo-Facial Surgery. 2012;**40**:321-328

[25] Hehn J, Schwenk T, Striegel M, et al. The effect of PRF (platelet-rich fibrin) inserted with a split-flap technique on soft tissue thickening and initial marginal bone loss around implants: Results of a randomized, controlled clinical trial. International Journal of Implant Dentistry. 2016;**2**(1):13

[26] Borgonovo AE, Berardinelli FV, Favale M, et al. Surgical options in oroantral fistula treatment. The Open Dentistry Journal. 2012;**6**:94-98

[27] Del Fabbro M, Gallesio G, Mozzati M. Autologous platelet concentrates for bisphosphonate-related osteonecrosis of the jaw treatment and prevention. A systematic review of the literature. European Journal of Cancer. 2015;**51**(1):62-74

Chapter 8

Rapid Cytoreduction by Plateletapheresis in the Treatment of Thrombocythemia

Bela Balint, Mirjana Pavlovic and Milena Todorovic

Abstract

The objective of this chapter is to provide a systematic overview of current knowledge regarding therapeutic apheresis—primarily therapeutic plateletapheresis (TP)—and to summarize evidence-based practical approaches related to cytapheresis treatment of "hyperthrombocytosis" or "extreme thrombocytosis" (ETC). Our results of platelet (Plt) quantitative/qualitative analyses and evaluation of efficacy of apheresis systems/devices—on the basis of Plt removal and in vivo Plt depletion—will be presented. Our preclinical researches confirmed that in Plt concentrates, the initial ratio of discoid shapes was 70%, spherical 20%, and less valuable (dendritic/balloonized) shapes 10%—with morphological score of platelets (MSP = 300–400). After storage, the ratio of discoid and spherical shapes was decreased, while the less valuable ones progressively increased (MSP = 200). Electron microscopy has shown discoid shapes with typical ultrastructural properties. Spherical shapes with reduced electron density and peripheral location of granules/organelles were detected. Also, dendritic shapes with cytoskeletal "rearrangement," membrane system integrity damages, and pseudopodia formations were documented. Our clinical study demonstrated that TP was useful in ETC treatment and should help prevention of "thrombo-hemorrhagic" events—until chemotherapy, antiplatelet drugs, and other medication take effect. During TP treatment, Plt count and morphology/ultrastructure were examined. Plt functions by multiplate analyzer were evaluated. We concluded that intensive TP was an effective, safe, and rapid cytoreductive treatment for ET.

Keywords: platelet, morphology-ultrastructure, thrombocythemia, apheresis, plateletapheresis

1. Introduction

Besides supportive treatment using whole blood, red blood cell (RBC), platelet (Plt), and plasma transfusions, hemotherapy, in an extensive viewpoint, involves also performing different apheresis procedures to remove or replace altered or overproduced and accumulated blood constituents (cell and plasma contents, pathogens, metabolic products, etc.). Hemotherapy also involves collection and ex vivo manipulation of stem cells (SCs) with a goal of reconstitution of hematopoiesis in performing conventional transplantations and for use in the fields of regenerative medicine.

Thus, the basic aim of hemotherapy is reconstitution and maintenance of a "hemobiological" (hematological, immunological, hemorheological) equilibrium or homeostasis by the recovery of lacking blood components, as well as through the use of apheresis and other "hemo-modifying" procedures. By doing this, the circulating blood volume is reestablished, but also the "blood oxiform function" (capacity for oxygen binding/transport), hemostasis, and activity of "immune response mediators" are restored. The rationalized and well-timed hemotherapy is contributing to increased survival of patients with some "life-threatening" conditions.

During supportive hemotherapy of patients with potential beneficial clinical response, it is essential to perform the most rational and "well-timed" intervention in the most appropriate manner—that is determining the most efficient and approachable therapeutic procedures and techniques. Also, it is imperative to optimize specific indications of blood component therapy with rationalized thresholds.

Also, the most sensitive point during planning and performing therapeutic apheresis (TAph), especially cytapheresis, is determination of the most optimal approach to the treatment or, even better, the "time-frequency-intensity continuum" of the procedure. The introduction of cytapheresis treatment requires determination of total blood cell quantity within the organism, sometimes even determination of the ratio between cells in intravascular versus extravascular space. The evaluation of hematopoietic potential and the reserve in bone marrow, as well as calculation of the intensity of cell production ("cytogenesis-kinetics") in the peri-apheresis period, is important too.

Before evaluation of results of TP in the treatment of our patients, several principles and standards of the apheresis treatment will be presented. Thus, this chapter will treat briefly general information related to apheresis treatment [without therapeutic plasma exchange (TPE)] with specific emphasis upon cytoreductive potential or thrombocytodepletion using cytapheresis treatment and lastly with presentation of our own results using TP procedures by comparison of the efficiency of different generations and types of blood cell separators.

While practically all details on the ultrastructure of Plts are already identified and explained, some morphological and functional properties of these blood elements will be very concisely recapitulated. Therefore, prior to presenting of our own data of Plt fundamental researches and in clinical study setting, some elementary morphological/ultrastructural and functional characteristics of Plts will be given.

2. Apheresis/cytapheresis: a systematic overview

Apheresis or hemapheresis is a "hemomodulatory" procedure in which blood components (cells or plasma) are collected from donors (the so-called apheresis donation) or specific quantitatively or qualitatively altered constituents/factors are removed from the bloodstream, while the remainder (seldom combined with normal blood cells) is returned to the patient's circulation (named as TAph) [1–4].

The basic goal of TAph procedures is to reduce the "patient's load" of specific substances, responsible for the development of disease, to the levels that will allow clinical improvement [4–6]. TAph treatment should be done regularly or always combined with "immunomodulatory" (e.g., immunochemotherapy) and other palliative (drug) therapies. Based upon the criterion of what is the removed blood constituent—with specific "pathogenic factors or substrate"—it is possible to classify the TAph procedures into TPE and therapeutic cytapheresis (TC) [5–15].

TPE is a replacement of the patient's plasma with an adequate substitution fluid (albumin in saline) or with normal (allogeneic) or modified (immunoadsorbed autologous) plasma. The use of "selective" TPE results in reduced risk of side effects (immunomodulation or virus transmission) since "purified" autologous plasma is used as a replacement fluid [11, 12].

TC is an effective "disease-modifying" therapeutic approach in which elevated or altered cells are removed from the blood, while the residue is returned into the circulation. The basic goal of the evidence-based TC procedure is a reduction of the "patient's load" with overproduced cell quantity or functionally altered cells, responsible for disease development [16–18]. TC procedures combined with immunochemotherapy and other medications result in "hemobiological" and overall clinical recovery. Thus, the main objective of TC procedures (symptomatic or prophylactic) is to reduce the patient's excessive cellular count—ex vivo Plt removal with in vivo Plt depletion (Plt removal/depletion)—and to alleviate symptoms created by these cells [1, 2, 18].

The goal of novel TC procedures—such as photopheresis, granulocyte adsorptive apheresis, and RBC exchange (RBCX)—is the substitution of abnormal or "disease-responsible" cells, as well as ex vivo modification of "immunocompetent cells" prior to reinfusion, getting frequently long-term beneficial clinical effects. RBCX procedures are indicated for treatment of patients with irregular RBCs and consecutive hemorheological abnormality with anemia and other hemobiological disturbances [6, 7, 16].

Using standard TC procedures, the excessive cellular quantity is removed from the patient's bloodstream with intention to reduce total blood cell count in organism and to prevent the symptoms and signs caused by various "cytemias." According to blood cells removed, TCs can be categorized into therapeutic leukapheresis (TL), erythrocytapheresis (TE), and TP [6, 16–18].

The aim of TLs is white blood cell (WBC) count reduction and reverse of the "hyperleukocytosis-leukostasis" syndrome, regression of organomegaly and lymphadenomegaly, and improvement in general clinical status [2, 15–17]. TLs represent a useful "cytoreductive" therapy during "leukostasis crisis" when the number of circulating WBC $\geq 150 \times 10^9$/L. In the treatment of patients with excessive number (500×10^9/L or more) of leukemic cells, their removal by TL would have been adequate for "partial excision" of tumor mass even 1000–1500 g of "total cell mass" in removed cell suspension—that is in the apheresis product (AphP) [7, 17].

The objective of TE treatment is to obtain a hemorheological advance. Besides removal of the excess of RBCs from patient's circulation, with TE it is possible to achieve hematological and hemorheological recovery (cellular hyperviscosity reduction) with consecutive improvement of tissue perfusion and oxygenation [2, 17].

Thrombocytapheresis or TP is useful in the treatment of patients with thrombocythemia and symptomatic thrombocytosis until chemotherapy, antiplatelet drugs, and other medications takes effect. The purpose of TP might be determined as a method for rapid reestablishment of Plt overproduction, following prevention of "thrombo-hemorrhagic" events. Thromboembolism is more frequent than bleeding crisis, and the arterial thromboses (followed by cerebrovascular, coronary, or other occlusive episodes) happen more often than venous events [17–22].

The use of TP treatment is indicated when the Plt number $\geq 1500 \times 10^9$/L ("hyperthrombocytosis" or "extreme thrombocytosis"), although the degree of thrombocytosis and the complexity of clinical status are not always causally associated. However, Plt quantification is undoubtedly helpful in predicting patient's hemostatic risk [22–29]. The details of TP treatment will be presented and discussed further below.

Concisely, the goal of our original and innovative "multi-manner" TAph is to offer the possibility of rapid returning from a "life-threatening" (irreversible) emergency—altering simultaneously more than one blood constituent(s)—to the "district" of unbalanced, but reversible clinical condition, with a possibility for recovery of patient's hematological and overall clinical homeostasis [13–15].

It is important to highlight and emphasize that TAph is an aggressive method, by which there is a direct intervention in patient's circulation; therefore it might be followed by side effects and complications. In order to minimize them, different criteria for selection of blood component donors and patients have to be applied. The analysis of laboratory parameters (blood cell count and biochemistry examination, screening testing for coagulation, etc.) is required too, along with clinical follow-up of the patients for which the TAph treatment is predicted.

The use of TAph procedures is not possible without (1) updated personal knowledge in hematology, immunology, biochemistry, and hemorheology; (2) rationalized education for treatment of acute "life-threatening" conditions; and (3) clinical experience related to working with extracorporeal circulation and cardiopulmonary reanimation. For that reason, TAph is applied only in medical centers with tertiary level of health protection and in countries with developed medical service.

3. Platelet morphology, ultrastructure, and functionality

Following contact with subendothelial structures at the sites of vascular injury, the "connecting" of Plts (adhesion) will occur in the presence of the von Willebrand factor (vWF). Then Plts are activated—a process which involves a number of "cell surface receptors" for agonists (ADP, thrombin, thromboxane A2 (TXA2), collagen, etc.), as well as specific signal transduction pathways. Activation process increases the intracellular concentration of calcium ions—by releasing from specific cytoskeleton structures and calcium influx across cell membrane. Elevated calcium concentration results in ultrastructural (morphological) and functional Plt alterations [17, 30–33].

After activation, Plts change their morphology—from discoid into spherical (seldom to dendritic) shapes—as a consequence of reorganization of the membrane-system and cytoskeleton network ("shape change reaction"). Specific granules are centralized intracellularly, and their contents are discharged into the open canalicular system and then to the outside of the cell ("releasing reaction"). The alteration in shape is the beginning of a series of reactions during Plt response to the action of agonists. Thus, discoid Plts transform into spherical or dendritic shapes, with pseudopod formation. These changes are practically prompt and do not require the presence of large amounts of extracellular calcium ions and fibrinogen. This "early" stage of Plt response is conditioned by polymerization of actin and by disconnecting of some marginal microtubule bonds. Elevated intracellular calcium concentration stimulates the phospholipase A2 (which releases arachidonic acid from membrane phospholipids). Arachidonic acid is then metabolized lastly into TXA2—a potent Plt activator [6, 31–37].

The "shape change reaction" allows the Plts to interact with one another ("clumping") in the blood to form aggregates. Therefore, the phenomenon of the Plt aggregation involves sequences of "platelet-to-platelet" interactions. This process requires the presence of extracellular calcium ions, as well as fibrinogen, with very close Plt surface contact [18, 30–32].

In a few words, Plt ultrastructure consists of three structural/functional zones: "peripheral zone," "sol-gel zone," and "organelle zone," as well as the "membrane-system."

The "peripheral zone" is responsible for Plt integrity and adhesion/aggregation—mediated by receptors and ligands (agonists). This zone consists of glycocalyx (external region), cytoskeleton or submembrane area, and cell membrane. Also, it incorporates various absorbed coagulation factors and receptors for thrombin, ADP, collagen, vWF, thrombospondin, TXA2, prostacyclin, fibrinogen, fibronectin, etc. [18, 31–33].

The "sol-gel zone" is a viscous cytoskeleton matrix, which contains an open canalicular system or microtubules, microfilaments, some vesicles, and secretory organelles. Microtubules support the membrane contractile cytoskeleton and maintain a discoid Plt shape. In the microfilaments actin, myosin, actomyosin, and other substances were detected. Actin microfilaments in the "sol-gel zone" form actin cytoskeleton (matrix for organelles), while actomyosin is responsible for Plt contraction and may promote granule secretion [6, 32–35].

The "organelle zone" includes alpha-granules, dense granules, and lysosomal granules, glycogen particles, mitochondria, and others. Alpha-granules (50–80 per Plt) consist of vWF, fibronectin, platelet factor 4, beta-thromboglobulin, thrombospondin, platelet-derived growth factor (PDGF), and fibrinogen. The dense granules (3–8 per Plt) include adenine nucleotides, serotonin, pyrophosphate, calcium, and magnesium ions. The lysosomal granules (up to 3 per Plt) contain acid hydrolases, cathepsin D and E, lysosomal-associated membrane protein (LAMP)-2 and some lipolytic enzymes, and CD63. Finally, the mitochondria are also in the "organelle zone" and provide Plt energy requirements. They are also a deposit and an important discharger of calcium ions [17, 30–35].

The open canalicular system of "membrane-system" can be used for transport of some plasma components (e.g., fibrinogen) to alpha-granules and for release of granular contents following Plt activation. The dense tubular system operates as a depot of calcium ions, and it is active during Plt contractions. Prostaglandin synthesis is discovered in these systems also after Plt activation [6, 30–32].

Generally, investigation of Plt functions is important for diagnosis of pathological conditions with qualitative Plt disorders—e.g., hemorrhage tendency or risk of interventional bleeding, especially in patients treated by antiplatelet drugs—also for determination of "storage lesions" (liquid-state conserved or cryopreserved cells), as well as for the establishment of the "critical threshold" for Plt supportive therapy [31–38].

Testing by Multiplate analyzer—a rapid assay of whole-blood impedance aggregometry—can be used to detect Plt dysfunction and for prediction of bleedings and/or some thrombotic events, also to resolve a personalized antiplatelet therapy (including monitoring of the treatment efficacy too), as well as to reduce of Plt transfusion needs. However, this testing does not give information concerning the Plt "shape change reaction" and the reversibility of aggregation and consequently cannot be used for diagnosis of some specific Plt function disorders [18]. The results of our own Plt functional examinations will be shown afterward.

In our initial cryobiological research studies [34–38], cells were investigated in the "buffy coat"-derived Plt concentrates (BC-PCs) intended for prophylactic or therapeutic transfusions. Recently, in our clinical study [18], Plts were evaluated before and following TP treatment in the peripheral blood of ET patients, as well as in the removed AphP.

In these studies, cells were quantified in the donors' blood samples and also in BC-PC units, as well as in the peripheral blood of patient before vs. after TP procedures and in the AphP using flow cytometry by Advia 2120 counter (Bayer, Germany). Morphological properties of Plts were examined by phase-contrast microscope (Polyvar, Reichert-Jung, Austria). The value of morphological score of platelets (MSP) was determined according to the percentage of different platelet shapes. Numerical valuation used for Plt shapes is as follows: balloonized = 0; dendritic = 1; spherical = 2; and discoid = 4. Ultrastructure of platelets was examined with electron microscope (Philips 201 C, Philips, The Netherlands). The Plt function by Multiplate analyzer (Dynabyte GmbH, Germany) was evaluated. Surface antigens were analyzed by flow cytometer Epics XL (Coulter, USA) using specific monoclonal antibodies [34–38].

Our preclinical studies [34–38] verified that in BC-PC units immediately after preparation ("unmanipulated" or unfrozen cells), the ratio of discoid shapes was about 60–80%, spherical approximately 15–25%, and functionally less valuable (dendritic or balloonized) shapes around 5–5%; then the value of MSP was between 300 and 400. During storage, the incidence of discoid and spherical shapes progressively decreased, while the less valuable ones increased. As a result, values of the MSP also reduced gradually. In BC-PC units stored in liquid state for up to 5 days, the MSP was approximately 200 [36].

The objective of our early cryobiological researches [34–42] was to determine an optimized cryopreservation protocol with minimized "thermal damages" (cryo-injury) of frozen/thawed cells—retaining maximum quantitative (cell count) and qualitative (cell functionality, morphology, and ultrastructure) recovery. Different cryopreservation protocols with controlled-rate ("microprocessor-controlled") or uncontrolled-rate ("dump-freezing" with no programmed cooling rate) freezing procedures, in combination with dimethyl sulfoxide (DMSO; final concentrations = 5, 6, and 10%), were compared [34–38].

Preclinical researches demonstrated that the frequency of discoid shapes in all Plt cryopreservation settings was lower (range: 39–58%) than in "unmanipulated"

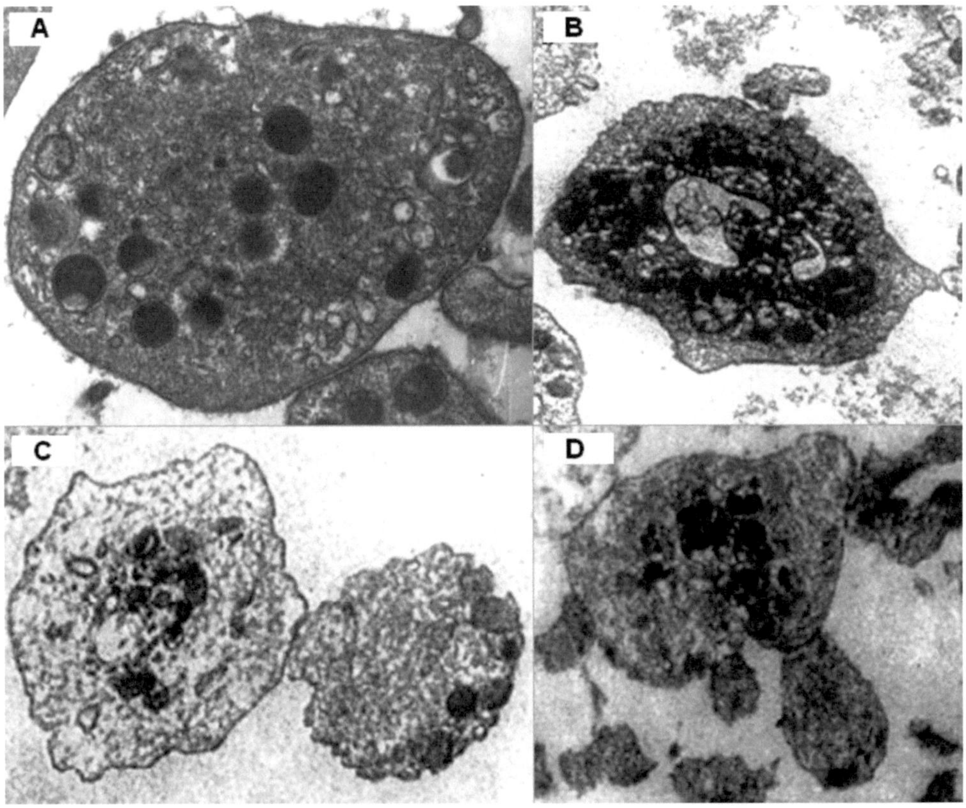

Figure 1.
Ultrastructure of Plts in cryobiological setting (our original data). Several discoid shapes had an unchanged "ultrastructural arrangement," with minor membrane injuries (A). Spherical shapes showed an inferior "electron density" of the cytoskeleton, somewhat inferior cell structure compared to discoid shapes—which was manifested by marginal location of alpha-granules, dense granules, lysosomal granules, and other organelles in the cytoskeleton (B and C). Several dendritic shapes with reduced quantity of alpha granules, numerous pseudopodia, and unclear cell edges were also confirmed (D). Sporadic balloonized shapes with critical membrane damages and multiple ruptures, as well as fragments of the destroyed Plts, were also detected.

(unfrozen) BC-PC group [34]. It is also confirmed that controlled-rate freezing provides higher percentage of discoid and spherical shapes—with resultant superior MSP value. In the opposite, a higher proportion of functionally less valuable shapes (dendritic or balloonized) was observed in uncontrolled-rate cryopreservation setting [35].

Electron microscopy in our Plt cryopreservation researches [35–38] had shown the occurrence of different Plt types—mainly discoid and spherical, as well as sporadic dendritic or balloonized shapes (as presented in **Figure 1**).

Finally, in our recent clinical study [18], the "shape ratio" and ultrastructure of Plts in peripheral blood of ET patients with ETC (treated by TP) were also investigated (Plt shapes are visualized in **Figure 2**).

A reduction of Plt function in patients with some disorders, also during liquid-state storage or following cryopreservation, is usually caused by stated ultrastructure "rearrangement" or "reorganization." The Plt membrane-system damage or destruction—with increased membrane permeability—can happen during the "phase change of the lipid." Finally, the "disarrangement" of receptors on the Plt membrane results in interruption of signal transduction pathways during activation process.

Our clinical study [18] also confirmed that the baseline aggregability of Plts in the patient's venous blood sample was 918 AU*min (normal = 923–1509 AU × min) by TRAP test (AU = aggregation unit; 10 AU × min is equal to one AUC [area under the curve] value; TRAP = thrombin receptor activating peptide). The results obtained indicated a merely Plt dysfunction (decreased activity) in the bloodstream of ET patient with ETC.

In our earlier research studies [34, 36–38] of "unmanipulated" and cryopreserved BC-PCs, various surface antigens were analyzed by flow cytometry using

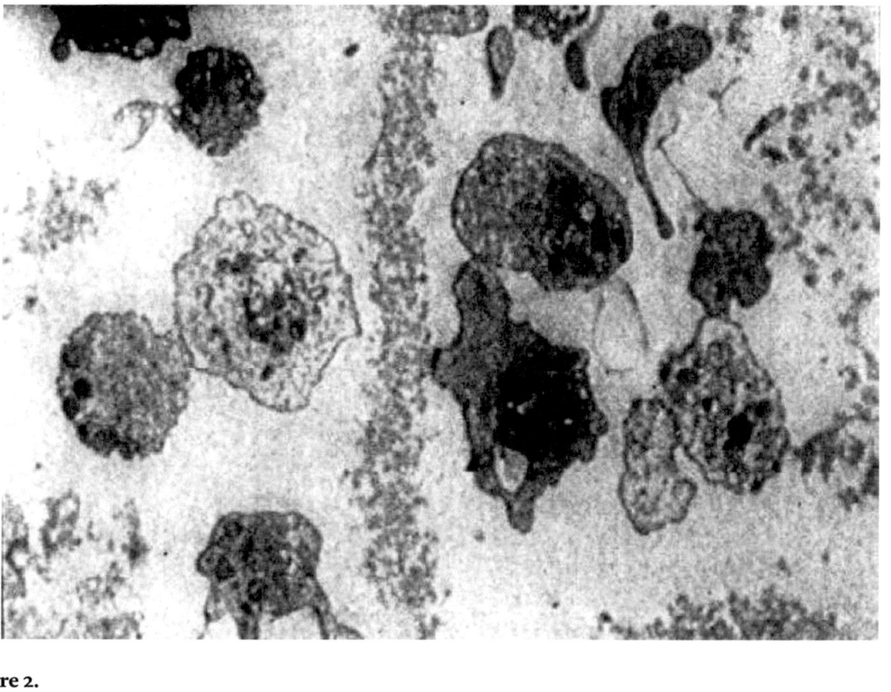

Figure 2.
Electron microscopy of Plt shapes in ET patient bloodstream (our own data). Discoid shapes had typical ultrastructural properties with intact microtubules and open canalicular system, without damage of the membrane-system integrity. Spherical shapes by peripheral location of granules/organelles in the cytoskeleton were also recognized (with resulting Plt dysfunction). A small number of dendritic shapes with cytoskeletal "rearrangement," membrane-system damages and pseudopod formations were also recognized.

monoclonal antibodies (MoAb) anti-GPIX/CD42a, anti-GPIIb-IIIa/CD41, anti-GPIbα/CD42b, anti-GP140/P-selectin/CD62p, anti-GP53/CD63, and anti-GPIV/CD36. Plts were incubated with fluorescein isothiocyanate or phycoerythrin-conjugated MoAb and then tested by flow cytometer.

Plts were evaluated on the basis of their characteristic cell size ("linear forward scatter") and granularity ("log side scatter") too [34]. Anionic phospholipid expression was tested using annexin V binding assay by flow cytometer. Cell viability was assessed by the hypotonic shock response test [34–36]. In the supernatant of BC-PCs, the levels of transforming growth factor β (TGF-β), soluble P-selectin, soluble annexin, platelet factor 4 (PF4), and β-thromboglobulin (β-TG) were tested by enzyme-linked immunosorbent assays [34, 36].

These studies verified that the GPIb/CD42b expression was reduced in all cryopreserved BC-PC groups in opposition to "unmanipulated" or unfrozen (control) group [36, 37]. The expression of GP140/CD62p was in all cryopreservation groups higher than in control [36]. It was concluded that evaluation of expression of activation markers on Plt surface and flow cytometric analysis of Plt subpopulations could be a helpful approach for the quality control of liquid-stored or cryopreserved BC-PCs [34].

4. Thrombocythemia and therapeutic thrombocytapheresis

Before evaluation of cytapheresis cytoreductive efficacy in our studies—using different apheresis systems in critical "clinicopathological" conditions with ETC, in which apheresis is beneficial, required, or sometimes might result in rescue of patient with acute "life-threatening" hazardous situations—common thrombocytosis-related data will be summarized.

4.1 Symptomatic thrombocytosis and thrombocythemia

The "critical threshold" for clinical significance of increased Plt count is variable from patient to patient—but, usually, conditions with Plt $\geq 450 \times 10^9$/L is designated as thrombocytosis. If thrombocytosis—typically with Plts $\geq 1500 \times 10^9$/L—appears within different myeloproliferative disorders, regularly we would talk about essential thrombocythemia (ET) [43–47].

Therefore, conditions with thrombocytosis and/or ETC can be classified into (1) primary disorder or ET, which is the consequence of clonal disease of the pluripotent hematopoietic stem cell, and (2) reactive or secondary thrombocytoses, which might evolve within patients with malignancies, chronic inflammatory processes, after splenectomy, etc. The existence of ETC induces an acquired "thrombo-hemorrhagic" diathesis, occasionally with potentially fatal vascular consequences [23–29, 43–50].

The entity and nature of ET is like polycythemia vera (PV), primary myelofibrosis (PMF), and myeloproliferative neoplasm (MPN) [46]. Due to the disease course, some patients with ET or PV might progress into a PMF-like post-ET or post-PV myelofibrosis [47]. Namely, all three MPN share three commonly exclusive "driver" mutations: JAK2, CALR, and MPL3. The most frequent is JAK2V617F, which originates in approximately 99% of patients with PV, 55% with ET, and 65% with PMF [48]. Concerning other driver mutations, 15–30% patients are CALR mutated, and 4–8% are MPL mutated, while 10–20% of the patients might not express any one of the three mutations, so they can be triple-negative [49].

Hemostatic mechanisms, by which thrombocytosis lead to "thrombo-hemorrhagic" events, are not completely explained. Different hemostatic defects and cell abnormalities are described—such as altered Plt aggregability, intracellular accumulation of specific active substances, decreased activity of cofactors von Willebrand ristocetin and multimers of vWF with high molecular mass, etc. Diagnosis of thrombocythemia is established on the basis of the existence of the following criteria of the World Health Organization (WHO): (1) persistent Plt count $\geq 450 \times 10^9$/L in the blood; (2) the presence of one of the three mentioned driver mutations or in their absence the exclusion of other causes of thrombocytosis (reactive and clonal); and (3) bone marrow morphologic evaluation, especially for discriminating ET from pre-fibrotic PMF and "masked" PV [50].

In ET or with difficult cases of secondary or reactive thrombocytosis (some carcinomas or asplenia), characteristic manifestations are headaches, vertigo, transitory visual disturbances, modest or intense chest or abdomen pain, as well as acrocyanosis, paresthesia, and other disturbances (e.g., priapism). In these patients, it is common manifestation of "thrombo-hemorrhagic" syndrome—thrombosis, with successive hemorrhages (epistaxis and other minor bleedings), as well as increased occurrence of first trimester miscarriage. However, in thrombocytosis or thrombocythemia, it is not undoubtedly proven that there is a correlation between the degree of increased Plt count and complexity of patients' clinical condition.

In therapy of ET patients, the main reason for treatment is to prevent "thrombo-hemorrhagic" episodes, and from this point of view, none of the most recent drugs have been shown to be superior among the traditional drugs, such as hydroxyurea. Three major risk factors for thrombosis are history of thrombotic events, JAK2/MPL mutations, and advanced age. In order to classify of ET patients—according to risk factors—there are four categories: "very-low-risk" (absence of all risk factors); "low-risk" (presence of JAK2/MPL mutations); "intermediate-risk" (presence of advanced age ≥60 years); and "high-risk" (presence of thrombotic events or occurrence of both JAK2/MPL mutations and advanced age) [48].

Patients with "very low risk" can be only observed or can take once-daily aspirin, while patients with "low-risk" disease take once- or twice-daily aspirin. In intermediate-risk setting, hydroxyurea and once-daily aspirin is therapy of choice, while in "high-risk" disease with arterial thrombosis, hydroxyurea and twice-daily aspirin is standard therapy. High-risk patients with venous thrombosis take hydroxyurea plus systemic anticoagulation, and when the JAK2/MPL mutation or cardiovascular risk factors are present, addition of aspirin is needed. Thus, therapy of these disorders includes the application of chemotherapy, antiaggregation, and other medications, as well as the use of TP procedures, when rapid reduction in Plt count is urgently needed (apheresis Plt depletion). Hydroxyurea is considered the drug of the first line for "medicamentous cytoreductive therapy." Second-line therapeutic drugs are pegylated interferon-α (IFN-α), busulfan, anagrelide, and pipobroman [48].

4.2 Therapeutic thrombocytapheresis

Generally, in ET patients by ETC (Plt $\geq 1500 \times 10^9$/L) and with altered Plt morphology/aggregability (Plt dysfunction) and immature reticulated Plts, the risk of "thrombo-hemorrhagic" event incidence is enlarged up to 50–60% [18, 22].

In the treatment of symptomatic ET—when low-dose aspirin or other antiplatelet and high-dose chemotherapies are without adequate response or contraindicated, as in pregnancy—cytoreduction by TP is beneficial or even essential [1, 2, 51]. The evidence-based clinical guideline for therapy of asymptomatic ET (e.g., exact cytapheresis threshold, initial and target Plt count, etc.) is not yet established. The treatment of ET in pregnancy is still mainly individualized [51]. The first TC procedure in our apheresis center was performed in 1971 for treatment of pregnant women with "hyperleukocytosis-leukostasis" syndrome [17].

Generally, the most critical step in TC therapy of patients with different types of "cythemia" (blood cell overproduction) is to define the optimal timing, frequency, as well as the intensity and duration of treatment. The use of TC procedures requires determination of excessive cell category and total cell quantity and to resolve also the predominant cell divisions (including intravascular vs. extravascular portions), as well as to establish and compensate for "peri-apheresis" intensity of novel blood cell production [15–17].

The goal of TP (in combination with described medication) in therapy of ET patients with clinical symptoms is to minimize of consequences of ETC. TC procedures performed using various devices (blood cell separators) operate in a wholly automatic manner—thus allowing "intra-apheresis" cell recruitment or refreshment and also minimizing the manual or individual (physician) variations and differences during treatment [16–18].

Therefore, TP procedures are used to obtain an effective and rapid cytoreduction in patients with ETC, in order to prevent or reduce "thrombo-hemorrhagic" events. The application of TP could be a life-saving procedure for selected ET patients with "life-threatening" clinical situation—due to acute episode of ETC. This treatment results in a dramatic Plt count fall and following clinical improvement [2, 22, 43–45]. As well, TC procedures are helpful in the treatment of patients with symptomatic reactive or secondary thrombocytosis (all together with correction of the cause of ETC).

However, the definitive decision to apply of TP treatment should be "individualized" on the basis of clinical scenario and intensity of ETC, as well as patients' risk profile [29]. In this context, the efficacy of TP is dependent on the initial Plt count, kinetics of cell production, and the volume of whole blood processed, but also on the efficacy of different apheresis devices.

In our clinical settings [8, 16–18], typically antecubital (seldom subclavian or jugular) veins were used for vascular access. During TP procedures patients were anticoagulated by acid-citrate-dextrose formula B solution (ACD-B; USP; with 1.8% citrate concentration) or ACD-A solution (ACD-A; USP; with 2.2% citrate concentration). A systemic heparinization was performed only in singular TP procedures. All patients tolerated intensive TP treatments well without any adverse effects.

The aim of our latest clinical study [18] was to evaluate the cytoreductive potential of the Spectra Optia/IDL System, based upon ex vivo Plt removal and in vivo Plt depletion (Plt removal/depletion) efficacy in the treatment of a 68-year-old female patient suffering from symptomatic ET (with headaches, vertigo, visual disturbances, and paresthesia). Modifications of manufacturer's original apheresis protocol included the collection preference and inlet flow correction (altered collection speed), as well as an increase of the "target cell suspension" volume [18]. Plt removal/depletion efficacy obtained in this study was compared to our earlier results (historical database) [8, 17] and the recent literature data [23–25] for different devices.

To the best of our knowledge, our clinical study [18] was only the second published clinical evaluation of the efficacy and safety of TP treatment using Spectra Optia/IDL System.

In earlier studies [8, 17], using apheresis devices of the first and the second generations (Haemonetics M-30, IBM 2997, and Cobe Spectra), TP procedures were performed typically every second day (rarely every day)—combined with standard immunochemotherapy and other medications. Namely, in the treatment of our comparable ET patients (n = 20; procedures = 126; historical database), applying mentioned apheresis devices, by one single TP procedure, it was possible to remove approximately 3×10^{12} Plt/L in cell suspension approximately

800–1300 mL or total 7–10 × 10^{12} Plts by one "whole therapeutic cure" (typically 5 single TPs; range = 3–11).

Opposite, in the treatment of recent ET patient with ETC-associated clinical emergency—using apheresis device of the newest generation and an intensive single TP treatment—the quantity of removed Plts was 7.5 × 10^{12} in the cell suspension (volume = 1150 mL) [18]. As replacement fluid, albumin in saline was used. There were no side effects due to intensive TP. Consequently, by Spectra Optia/IDL System using one TP, it was possible achieve similar therapeutic effect to that one which was realized after the application of "whole therapeutic cure" with devices of earlier generations.

In Cobe Spectra group (historical database) of our patients [17], the in vivo Plt depletion was approximately threefold lower. Only the use of "whole therapeutic cure" resulted in a satisfactory in vivo Plt fall (mean = 68 ± 14%; range = 55–85%). Recent literature data [28] are comparable—precisely, a 67% decrease in circulating Plt count (1553 × 10^9–513 × 10^9/L) was reported, but after two TP procedures. Contrary to our recent study [18], a significant Plt count depletion (from 2330 × 10^9 to 633 × 10^9/L) and excellent in vivo Plt fall (72.8%) were realized by one single TP procedure—followed by clinical advances and prevention of potential "thrombo-hemorrhagic" events (e.g., cerebrovascular accident—stroke).

The in vivo Plt depletion intensity in our current study (72.8%) was also superior among the most recent literature data [23–25]. Exactly, using a single TP by different apheresis devices, such as Haemonetics-MCS+ (therapy of low-body-mass child; Plts = 3072 × 10^9/L) [52], CS-3000 Plus and Cobe Spectra (management of hemato-oncological patients) [25], as well as Spectra Optia apheresis system (treatment of ET patient; manufacturer's protocol used) [24], the post-apheresis reductions of in vivo Plt depletion averaged only 37, 38, and 56%, respectively.

5. Conclusion

The purpose of TAph treatment is to reduce the "patient's load" of specific substances, responsible for the development/progress of disease, to the levels that will allow clinical recovery. Using TC procedures, the excessive cellular quantity is removed from the patient's bloodstream with the intention to reduce total blood cell count in the body and to prevent the symptoms caused by various "cytemias." TAph is an aggressive method; therefore it might be followed by side effects and complications. Thus, the use of TAph procedures is not possible without personal education for treatment of acute "life-threatening" conditions, as well as clinical experience related to working with extracorporeal circulation and cardiopulmonary reanimation.

After activation, Plts change their morphology—from discoid into spherical (seldom to dendritic) shapes—as a consequence of reorganization of the membrane-system and cytoskeleton network ("shape change reaction"). Specific granules are centralized intracellularly, and their contents are discharged into the open canalicular system and then to the outside of the cell ("releasing reaction"). The alteration in shape represents the beginning of a series of reactions during Plt response to the action of agonists. Investigation of Plt functions is important for diagnosis of pathological conditions with qualitative Plt disorders—e.g., hemorrhage tendency or risk of interventional bleeding, especially in patients treated by antiplatelet drugs— and also for determination of "storage lesions" (liquid-state conserved or cryopreserved cells), as well as for the establishment of the "critical threshold" for Plt supportive therapy. The evaluation of membrane antigens and flow cytometric analysis of Plt subpopulations could be helpful for the quality control of stored BC-PCs.

Based on the literature data and our own results, it is possible to conclude that intensive TP is an effective and safe procedure—even in long-term application (repeated or intermittent procedures) over a period of a few weeks/months—for patients with ET or secondary thrombocytoses and "life-threatening" ETC.

The application of a single TP procedure in the treatment of recent ET patient by Spectra Optia/IDL System resulted in undoubtedly superior Plt removal/depletion efficacy (for both normal and altered cells) when compared to our earlier study (Cobe Spectra) and literature data for CS-3000 Plus or Cobe Spectra, even for the Spectra Optia (when the manufacturer's original protocol was used). Although TPs represent an effective cytoreductive therapy with high level of Plt removal/depletion potential, it cannot influence clinical remission.

Author details

Bela Balint[1,2]*, Mirjana Pavlovic[3] and Milena Todorovic[4]

1 Department of Transfusion Medicine, Dedinje Cardiovascular Institute, Belgrade, Serbia

2 Department of Medical Sciences, Serbian Academy of Sciences and Arts, Belgrade, Serbia

3 Department of Computer and Electrical Engineering and Computer Science, Florida Atlantic University, Boca Raton, Florida, USA

4 Clinic for Hematology, Clinical Center of Serbia, Belgrade, Serbia

*Address all correspondence to: balintbela52@yahoo.com

References

[1] Fung MK, Eder AF, Spitalnik SL, Westhoff CM, editors. Technical Manual. 19th ed. Bethesda: AABB Press; 2017

[2] Padmanabhan A, Connelly-Smith L, Aqui N, Balogun RA, Klingel R, Meyer E, et al. Guidelines on the use of therapeutic apheresis in clinical practice – Evidence-based approach from the Writing Committee of the American Society for Apheresis: The eighth special issue. Journal of Clinical Apheresis. 2019;**34**:171-354

[3] Balint B, Ljubenov M, Stamatovic D, Todorovic M, Pavlovic M, Ostojic G, et al. Stem cell harvesting protocol research in autologous transplantation setting: Large volume vs. conventional cytapheresis. Vojnosanitetski Pregled. 2008;**65**(7):545-551

[4] Balint B, Stanojevic I, Todorovic M, Stamatovic D, Pavlovic M, Vojvodic D. Relative frequency of immature $CD34^+$/$CD90^+$ subset in peripheral blood following mobilization correlates narrowly and inversely with the absolute count of harvested stem cells in multiple myeloma patients. Vojnosanitetski Pregled. 2017;**74**(11):1071-1077

[5] Boljevic D, Nikolic A, Rusovic C, Lakcevic J, Bojic M, Balint B. A promising innovative treatment for ST-elevation myocardial infarction: The use of C-reactive protein selective apheresis: Case report. Blood Purif. 28 February 2020:1-5. PMID: 32114573. DOI: 10.1159/000506176. [Online ahead of print]

[6] Klein HG, Anstee DJ, editors. Mollison's Blood Transfusion in Clinical Medicine. 12th ed. Oxford: Wiley and Blackwell; 2013

[7] McLeod BC. Apheresis: Principles and Practice. 3rd ed. Bethesda: AABB Press; 2010

[8] Balint B, National Apheresis Group. Apheresis in donor and therapeutic settings: Recruitments vs. possibilities – A multicenter study. Transfusion and Apheresis Science. 2005;**33**(2):229-237

[9] Radovic M, Balint B, Tiska-Rudman LJ, Milenkovic LJ, Taseski J. Application of therapeutic plasma exchange in hematology. Transfusion Science. 1994;**15**:481-485

[10] Radovic M, Balint B, Jovicic A. The use of therapeutic plasma exchange for treatment of acute polyradiculoneuropathy. Transfusion Science. 1995;**16**:167-171

[11] Balint B, Pavlovic M, Jevtic M, Hrvacevic R, Ostojic G, Ignjatovic LJ, et al. Simple "closed-circuit" group-specific immunoadsorption system for ABO-incompatible kidney transplants. Transfusion and Apheresis Science. 2007;**36**:225-233

[12] Balint B, Pavlovic M, Todorovic M, Jevtic M, Ristanovic E, Ignjatovic LJ. The use of simplified ex vivo immunoadsorption and "multi-manner" apheresis in ABO/H-mismatched kidney transplant setting – Phase II clinical study. Transfusion and Apheresis Science. 2010;**43**(2):141-148

[13] Todorovic M, Balint B, Suvajdzic N, Pavlovic M, Petrovic M, Jevtic M, et al. Triple-way therapeutic approach for paraganglioma-dependent erythrocytosis: Drugs plus surgery plus "multi-manner" apheresis. Medical Oncology. 2008;**25**(2):148-153

[14] Balint B, Todorovic-Balint M, Petakov M, Ostojic G, Vucetic D. Effectively "cross-bridged" hemostatic and blood screening test defects due to glycogenosis type 1 associated extremely hyperlipidemia. Transfusion and Apheresis Science. 2014;**50**:314-315

[15] Balint B, Todorovic-Balint M, Ljubenov M, Tarabar O, Pavlovic M, Kanjuh V. Apheretic "rescue-protocol" designed for treatment of CLL associated life-threatening hemolytic crisis. Transfusion and Apheresis Science. 2015;**52**(2):256-258

[16] Radovic M, Balint B, Milenkovic LJ, Tiska-Rudman LJ, Taseski J. Therapeutic leukapheresis. Transfusion Science. 1991;**12**:193-196

[17] Balint B, Ostojic G, Pavlovic M, Hrvacevic R, Pavlovic M, Tukic LJ, et al. Cytapheresis in the treatment of cell-affected blood disorders and abnormalities. Transfusion and Apheresis Science. 2006;**35**(1):25-31

[18] Balint B, Pavlovic M, Ostojic G, Vucetic D, Todorovic M. Improved cytoreductive potential of plateletapheresis in treatment of symptomatic thrombocytosis: A single center study. Vojnosanitetski Pregled. 2019;**76**(5):559-560

[19] Panlilio AL, Reiss RF. Therapeutic plateletpheresis in thombocythemia. Transfusion. 1979;**19**(2):147-152

[20] Belloni M, Fabris F, Ongaro G, Girolami A. Therapeutic thrombocytapheresis in a case of thrombocytosis. Transfusion. 1981;**21**(2):229-230

[21] Langer R, Gross D, Mansouri Taleghani B, Wiebecke D, Henrich HA. Therapeutic thrombocytapheresis – Effect on hemorheologic parameters. Beiträge zur Infusionstherapie und Transfusionsmedizin. 1997;**34**:265-272

[22] Adami R. Therapeutic thrombocytapheresis: A review of 132 patients. The International Journal of Artificial Organs. 1993;**16** (Suppl 5):183-184

[23] Koren-Michowitz M, Lavi N, Ellis MH, Vannucchi AM, Mesa R, Harrison CN. Management of extreme thrombocytosis in myeloproliferative neoplasms: An international physician survey. Annals of Hematology. 2017;**96**(1):87-92

[24] Almeida-Dias R, Garrote M, Cid J, Mustieles MJ, Alba C, Lozano M. Therapeutic thrombocytapheresis for extreme thrombocytosis after chemotherapy in essential thrombocytosis. Journal of Clinical Apheresis. 2019;**34**(4):503-506

[25] Prakash S, Hans R, Sharma RR, Malhotra P, Marwaha N. Therapeutic thrombocytapheresis for symptomatic thrombocytosis in hemato-oncology patients. Therapeutic Apheresis and Dialysis. 2018;**22**(1):93-95

[26] Schafer AI. Thrombocytosis and thrombocythemia. Blood Reviews. 2001;**15**(4):159-166

[27] Tefferi A, Barbui T. Polycythemia vera and essential thrombocythemia: 2015 update on diagnosis, risk-stratification and management. American Journal of Hematology. 2015;**90**(2):162-173

[28] Thakral B, Saluja K, Malhotra P, Sharma RR, Marwaha N, Varma S. Therapeutic plateletpheresis in a case of symptomatic thrombocytosis in chronic myeloid leukemia. Therapeutic Apheresis and Dialysis. 2004;**8**(6):497-499

[29] Boddu P, Falchi L, Hosing C, Newberry K, Bose P, Verstovsek S. The role of thrombocytapheresis in the contemporary management of hyperthrombocytosis in myeloproliferative neoplasms: A case-based review. Leukemia Research. 2017;**58**:14-22

[30] Tian J, Cheng LH, Cui X, Lei XX, Tang JB, Cheng B. Investigating the effect of age on platelet ultrastructure using transmission electron microscopy.

International Wound Journal. 2019;**16**(6):1457-1463

[31] Ponomareva AA, Nevzorova TA, Mordakhanova ER, Andrianova IA, Rauova L, Litvinov RI, et al. Intracellular origin and ultrastructure of platelet-derived microparticles. Journal of Thrombosis and Haemostasis. 2017;**15**(8):1655-1667

[32] Pokrovskaya ID, Yadav S, Rao A, McBride E, Kamykowski JA, Zhang G, et al. 3D ultrastructural analysis of α-granule, dense granule, mitochondria, and canalicular system arrangement in resting human platelets. Research and Practice in Thrombosis and Haemostasis. 2019;**4**(1):72-85

[33] Pokrovskaya ID, Tobin M, Desai R, Joshi S, Kamykowski JA, Zhang G, et al. Canalicular system reorganization during mouse platelet activation as revealed by 3D ultrastructural analysis. Platelets. 31 January 2020:1-8. PMID: 32000578. DOI: 10.1080/09537104.2020.1719993. [Online ahead of print]

[34] Vucetic D, Ilic V, Vojvodic D, Subota V, Todorovic M, Balint B. Flow cytometry analysis of platelet populations: Usefulness for monitoring the storage lesion in pooled buffy-coat platelet concentrates. Blood Transfusion. 2018;**16**(1):83-92

[35] Balint B, Vucetic V, Trajkovic-Lakic Z, Petakov M, Bugarski M, Brajuskovic G, et al. Quantitative, functional, morphological and ultrastructural recovery of platelets as predictor for cryopreservation efficacy. Haematologia. 2002;**32**(4):363-376

[36] Balint B, Paunovic D, Vucetic D, Vojvodic D, Petakov M, Trkuljic M, et al. Controlled-rate vs. uncontrolled-rate freezing as predictors for platelet cryopreservation efficacy. Transfusion. 2006;**46**(2):230-235

[37] Balint B, Vucetic D, Vojvodic D, Brajuskovic G, Ivanovic Z, Trkuljic M, et al. Cell recovery, cryothermal micro-damages and surface antigen expression as predictors for cold-induced GPIbα/CD42b-cluster mediated platelet clearance after controlled-rate vs. uncontrolled-rate cryopreservation. Blood Banking and Transfusion Medicine. 2004;**2**(1):22-26

[38] Vucetic D, Balint B, Trajkovic-Lakic Z, Mandic-Radic S, Brajuskovic G, Mikovic D, et al. Cryopreserved vs. liquid-state stored platelets – Quantitative and qualitative comparative study. Blood Banking and Transfusion Medicine. 2003;**1**(2):81-88

[39] Balint B, Ivanovic Z, Petakov M, Taseski J, Jovcic G, Stojanovic N, et al. The cryopreservation protocol optimal for progenitor recovery is not optimal for preservation of MRA. Bone Marrow Transplantation. 1999;**23**:613-619

[40] Skoric D, Balint B, Petakov M, Sindjic M, Rodic P. Collection strategies and cryopreservation of umbilical cord blood. Transfusion Medicine. 2007;**17**(2):107-113

[41] Balint B, Stamatovic D, Todorovic M, Elez M, Vojvodic D, Pavlovic M, et al. Autologous transplant in aplastic anemia: Quantity of $CD34^+$/$CD90^+$ subset as the predictor of clinical outcome. Transfusion and Apheresis Science. 2011;**45**:137-141

[42] Todorovic Balint M, Jelicic J, Bila J, Balint B, Antic D, Trajkovic G, et al. Influence of applied $CD34^+$ cell dose on the survival of Hodgkin's lymphoma and multiple myeloma patients following autologous stem cell transplants. Vojnosanitetski pregled. 2020. Online-First CID: 81828431. DOI: 10.2298/VSP180808160T

[43] Rumi E, Cazzola M. How we treat essential thrombocythemia. Blood. 2016;**128**(20):2403-2414

[44] Aruch D, Mascarenhas J. Contemporary approach to essential thrombocythemia and polycythemia vera. Current Opinion in Hematology. 2016;**23**(2):150-160

[45] Kiladjian JJ, Besses C, Griesshammer M, Gugliotta L, Harrison C, Coll R, et al. Efficacy and safety of cytoreductive therapies in patients with essential thrombocythaemia aged >80 years: An interim analysis of the EXELS study. Clinical Drug Investigation. 2013;**33**(1):55-63

[46] Arber DA, Orazi A, Hasserjian R, Thiele J, Borowitz MJ, Le Beau MM, et al. The 2016 revision to the World Health Organization classification of myeloid neoplasms and acute leukemia. Blood. 2016;**127**:2391-2405

[47] Barosi G, Mesa RA, Thiele J, Cervantes F, Campbell PJ, Verstovsek S, et al. Proposed criteria for the diagnosis of post-polycythemia vera and post-essential thrombocythemia myelofibrosis: A consensus statement from the International Working Group for Myelofibrosis Research and Treatment. Leukemia. 2008;**22**(2):437-438

[48] Tefferi A, Alessandro M. Vannucchi A, Barbui T. Essential thrombocythemia treatment algorithm 2018. Blood Cancer Journal. 2018;**8**:2-6. DOI: 10.1038/s41408-017-0041-8

[49] Tefferi A, Pardanani A. Myeloproliferative neoplasms: A contemporary review. JAMA Oncology. 2015;**1**:97-105

[50] Tefferi A, Barbui T. Polycythemia vera and essential thrombocythemia: 2017 update on diagnosis, risk-stratification, and management. American Journal of Hematology. 2017;**9**:94-108

[51] Koh LP, Devendra K, Tien SL. Four pregnancies in two patients with essential thrombocythaemia – A case report. Annals of the Academy of Medicine, Singapore. 2002;**31**(3):353-356

[52] Singh A, Nityanand S, Sonker A, Kumar S. Successful use of the cell separator haemonetics multicomponent collection system+ for therapeutic thrombocytapheresis in a low body weight child of essential thrombocythemia. Asian Journal of Transfusion Science. 2015;**9**(2):207-209